中国水利成就系列

中国历代水利工程（二）

水利部国际经济技术合作交流中心　编译

中国水利水电出版社
www.waterpub.com.cn
·北京·

图书在版编目（CIP）数据

中国水利成就系列. 中国历代水利工程. 二 / 水利部国际经济技术合作交流中心编译. -- 北京 : 中国水利水电出版社, 2023.11
ISBN 978-7-5226-1900-2

Ⅰ. ①中… Ⅱ. ①水… Ⅲ. ①水利工程－水利史－中国 Ⅳ. ①TV-092

中国国家版本馆CIP数据核字(2023)第213586号

书　　名	中国水利成就系列 **中国历代水利工程（二）** ZHONGGUO LIDAI SHUILI GONGCHENG (ER)
作　　者	水利部国际经济技术合作交流中心　编译
出版发行	中国水利水电出版社 （北京市海淀区玉渊潭南路1号D座　100038） 网址：www.waterpub.com.cn E-mail:sales@mwr.gov.cn 电话：（010）68545888（营销中心）
经　　售	北京科水图书销售中心（零售） 电话：（010）68545874、63202643 全国各地新华书店和相关出版物销售网点
排　　版	中国水利水电出版社微机排版中心
印　　刷	河北鑫彩博图印刷有限公司
规　　格	184mm×260mm　16开本　11.25印张　171千字
版　　次	2023年11月第1版　2023年11月第1次印刷
印　　数	0001—1000册
定　　价	**98.00**元

《中国历代水利工程（二）》编委会

写在前面的话

水是生命之源、生产之要、生态之基。兴水利、除水害，古今中外，都是治国大事。从一定意义上说，治水即治国，治水之道是重要的治国之道。在中华民族5000多年的发展进程中，治水一直关乎民族生存、文明进步、国家强盛。水运连着国运，一部泱泱大国的治国史，也是一部百折不挠的治水史。

几千年来，从最大最复杂的地下灌溉工程坎儿井，到有“亚洲天池”美誉的丹江口水库，从福泽万民的世界灌溉工程遗产黄鞠灌溉工程，到被称为“世界第八大奇迹”的红旗渠，我国劳动人民在抗击洪水干旱和开发利用水资源方面创造了举世瞩目的成就。

纵览这些非凡的历代水利工程，您可以更加了解中国治水史和中华水文化，更加体会水利作为农业命脉和经济社会发展根本的重要意义，更加理解水安全是涉及国家长治久安的大事和实现中华民族伟大复兴的战略保障。

2023年11月

目录

CATALOGUE

一 世界上最大最复杂的地下灌溉工程：坎儿井

始于西汉时期，距今已有两千多年，目前仍在新疆地区使用的坎儿井（地下灌溉工程），是劳动人民为应对高蒸发量、少降雨量的干旱气候而发明出来的水利工程。

图 1　坎儿井入口处

坎儿井的主要设计原理是将地层中的潜流通过人工挖掘的暗渠逐步汇引到地面明渠浇灌农田。工程大体上包括竖井、暗渠、明渠和涝坝四部分。潜流主要由积雪融化渗入地下形成。人们在山前寻找潜流水，确定好引水路线后，先打若干竖井，再在地下打出暗渠，将其连接起来。水流完全利用自然坡降，最后到达灌溉区域。这样的灌溉方式最大程度地减少了沿途蒸发损失。在没有准确定位系统、劳动工具简陋的条件下，能够完全依靠人工地下作业实现工程目标，这确是一个伟大的创举。

今天，在中国新疆，现存有还在使用的坎儿井 1,700 多条，灌溉面积约 50 万亩，绝大部分集中在新疆吐鲁番地区。

人们常把坎儿井与长城、中国大运河并称为中国古代三大工程。

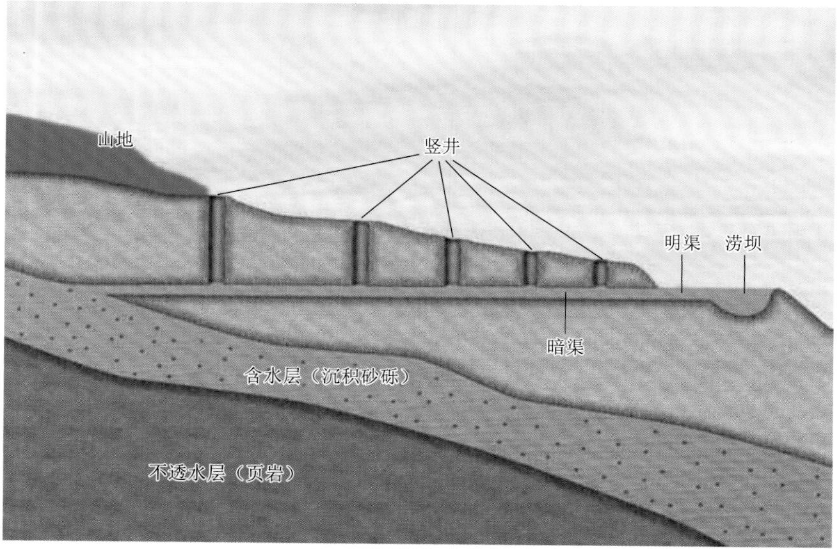

图 2　坎儿井结构示意图

二 历史最悠久的拱形坝体：通济堰

始建于南朝萧梁天监四年（505 年）、重建于宋开禧元年（1205 年），位于浙江省西南瓯江支流松阴溪上的通济堰，是世界历史最悠久、主要为灌溉服务的拱形拦河水利工程。

最初，通济堰为木质建筑，宋朝一辞官赋闲回乡的官员何澹奏请皇上重修通济堰，重修的通济堰改为石坝。坝长 275 米，宽 25 米，高 2.5 米。上游集雨面积约 3,150 平方公里，引水流量为 3 立方米 / 秒，整个坝体凸向上游约 120 度弧形。拱形拦水坝能更好地承受洪水冲力，其建设时间早于西班牙建于 16 世纪的爱尔其拱坝和意大利建于 1612 年的邦达尔多拱坝。

除了拱形坝体，这一水利工程还有诸多特点。一是在选择坝址时充分考虑自然高差，使配套灌溉渠道可以自流灌溉；二是大坝采用大松木作为坝基，铁水浇筑石坝，保证其基地不腐，石坝整体性能强，这也是大坝千年永固的重要原因之一；三是兼顾排沙和行船，大坝北端设有净宽 2 米两孔、深至坝底的排沙门，上游大水冲下来的沙石利用排沙门的急流自动排到大坝下面，还设有一座净宽 5 米的过船闸，保证古时的通航功能；四是建造了一座立体交叉石函引水桥，俗称“三洞桥”，让影响灌溉渠道安全的山溪洪水从桥面上通过，进入瓯江，灌溉渠水从桥下穿流，引洪灌溉两不误；五是通过 72 座闸以及连接的 3 条毛渠合理调节灌溉水量。自宋元至清，通济堰经过多次续建整修。

通济堰自古留有堰史、堰规，通济堰自创建以来，历代都比较重视修护和管理，有一套自成系统且完整的管理方法。管理办法中，最早有文字记载的“堰规”出现于北宋元祐七年（1092 年），现存最早的堰规是南宋乾道四年（1168 年）处州太守范成

图 1　通济堰俯视图

图 2　通济堰通水桥

大所制订的堰规，且独树一帜，科学而全面，堪为后人仿效，现存于大坝边上的詹南司马庙内。内容涵盖支流分水口规格、轮灌制度、岁修工役、工料摊派办法和工程巡视奖惩办法等。直到今天，部分制度仍在沿用。

今天，通济堰所在的地方风景秀丽。源头堰头村，群山环抱，古樟弥盖；“三洞桥”旁建有“文昌阁”；堰坝西侧建有一座詹南司马祠，俗称“龙庙”，祠内保存着宋、元、明、清及民国时期的碑刻 16 方，记录着历代修建情况及堰规、堰图等。汤显祖等文人墨客都曾留墨于此。

三　开凿千年，分荫万畦：黄鞠灌溉工程

黄鞠灌溉工程位于福建省宁德市蕉城区霍童镇，1,400 余年前由隋朝谏议大夫黄鞠率领黄氏家族在霍童溪沿岸兴建，是隋朝（581—618 年）系统最完备、技术水平最高的水利工程。实际上

图 1　黄鞠灌溉工程

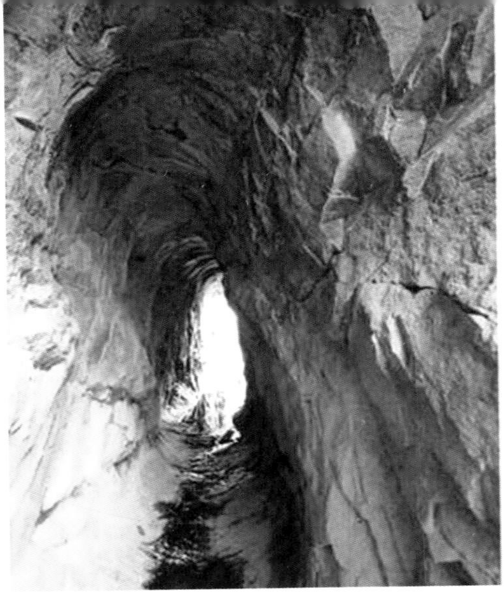

除了灌溉功能，它还发挥了提供生活供水功能，利用了水的势能，亦可称为一个小型的综合水利工程。2017年入选第四批世界灌溉工程遗产名录。

图2　琵琶洞

黄鞠灌溉工程分为左岸琵琶洞引水工程和右岸龙腰渠灌溉工程，总长十多公里，灌溉面积两万余亩。琵琶洞总长700多米，平均高2.41米、宽1米左右。沿霍童溪一侧设有排沙孔，便于清淤和排沙。

龙腰渠主干渠是全长5,000余米的明渠，宽1.51～2.72米，深0.95～3米。利用水位高差建有五级水碓，水车带动磨、碾、舂、筛用于加工农副产品和粮油。之后渠水又分为两支，一支灌

图3　龙腰渠

溉石桥洋千余亩良田，另一支供给村民生活用水，形成一整套完备的灌溉、水力、民生供水系统。

黄鞠灌溉工程的技术成就主要体现为，在无动力驱动时代，在恶劣的地理环境下，通过科学规划，利用地形高差，以堰坝拦水、明渠引水、穿隧洞引水，形成了引、输、蓄、灌、排的合理布局，实现了灌溉工程的多功能目标。采用火烧水激凿石工法，形成了合理的隧洞断面形状，增加了结构的稳定性。

黄鞠灌溉工程是民间自筹修建、政府管理的典范工程。兴利上千年，基本保持原貌，仍发挥着农业灌溉、生活供水、水力加工等综合功能，为当代和后代留下了灌溉文明的历史见证。

四

长江黄河一线牵：南水北调（中线）穿黄工程

南水北调（中线）穿黄工程位于河南省郑州市以西约 30 公里的孤柏嘴，是整个南水北调中线的标志性、控制性工程。"穿黄"即"穿越黄河"，南水北调中线工程将长江水自丹江口水库一路引向北方，黄河成为阻碍南水北去的天然屏障，为解决这一问

图 1　南水北调（中线）穿黄工程示意图

题，修建了穿黄工程这一人类历史上最大的穿越大江大河的水利工程。

穿黄工程，总长 19.3 公里，其中，明渠长 13.95 公里，隧洞段长 4.71 公里，建筑物长 0.65 公里。穿黄段采用“河下隧洞”方式：在黄河河床下 40 米深处，平行设置两条直径 7 米的隧洞横穿黄河；双隧洞的方案，可以保证在需要时加大流量，在某个洞发生状况时可保障供水不间断；隧洞采用“倒虹吸”原理，进水端（南岸）高于出水端（北岸），利用两端压差将水压向出水端；隧洞进口为斜井、出口为竖井，这种布置方案保证了隧洞具有良好的运行与检修条件，又节省工程投资。

图 2　穿黄工程进水端（南岸）

图 3　穿黄工程出水端（北岸）

考虑到隧洞不仅承受外部的水、土压力，同时承受洞内的流水压力；在设计上，隧洞洞壁采用 2 层钢筋混凝土衬砌，分别承受内、外压力；外层为厚 0.4 米拼装式管片结构衬砌，内层为厚 0.45 米钢筋混凝土预应力衬砌，两层衬砌之间采用透水垫层隔开。隧洞内布置许多监测设备和仪器，随时监控隧洞的运行状态。穿黄工程的抗震能力为 8 级，防洪能力按黄河 300 年一遇洪水设计，按 1000 年一遇洪水校核。

隧洞工程采用盾构方式掘进，盾构机直径 9 米，长 80 余米，总重量 1,100 吨。在施工过程中，克服了黄河河床游荡、复杂地质条件、砂土振动液化等一系列技术难题；第一次采用泥水平衡加压式盾构进行隧洞施工，第一次采用双层衬砌结构等多项首创技术。

图 4　隧洞内部

河床中存在的大量块石、古木等，地层成分的变化和顶部黄河水压力的变化，都对隧道的施工安全造成威胁。因而在整个隧洞的挖掘中，根据地质条件的变化，不断调整盾构机掘进参数、推力、扭矩、掘进速度和贯入度。

穿黄工程实现了汉江水与黄河的立体交叉，呈现出江河相会的壮美景观，体现了中华民族智慧与魄力。南水北调中线工程正式通水以来，实现 1,900 多天连续不间断安全供水，工程累计调水量已近 300 亿立方米，直接受益人口超过 1.2 亿人，经济、社会、生态等方面效益显著，已成为很多城市供水新的生命线。

五

汉中地区最早的农田灌溉水利工程：汉中三堰

位于陕西省汉中市的汉中三堰——山河堰、五门堰和杨填堰，是汉中盆地的重要灌溉工程，始建于西汉时期（公元前206—公元8年），距今已有2,000多年，是汉中地区最早的农田灌溉水利工程，它解决了汉中盆地汉江以北区域的灌溉难题，使汉中成为最早的"天府之国"之一。"汉中三堰"一直得到不断地整修

图1　五门堰渠首

维护，至今仍灌溉着 21.75 万亩农田。2017 年汉中三堰入选世界灌溉工程遗产名录。

汉中三堰之一山河堰，位于长江支流汉江的一级支流褒河上。传说是由汉代名将萧何和曹参主持修建。其堰头有三处，主要目的是截住褒河水用于灌溉农田。现代灌溉工程褒惠渠基本上是沿山河堰旧线修筑。1975 年，石门水库建成后，原山河堰所灌田亩尽纳入石门南干渠灌溉范围。

汉中三堰之二五门堰，位于陕西省城固县城北 15 公里处的湑水河右岸，湑水河也是汉江上游的一条支流。五门堰因渠首横列五洞进水故名五门堰，同样始建于汉代。初建时灌溉面积为 3,000 余亩，后经历代整修扩建，到清代时灌溉面积已达 5 万余亩，并具有一定的防洪功能，至今仍然具备一定的灌溉功能。

图 2　杨填堰闸门

汉中三堰之三杨填堰，位于陕西省城固县北约 10 公里处的湑水河中游段左岸，宋代以前称张良渠。1163—1165 年（南宋），杨从仪重新疏浚渠道，由此改名杨填堰（见《重刻汉中府志》），当时能够灌溉城固县农田七千亩，洋县农田一万八千亩。后来该堰渠几经洪水冲毁又多次重建。目前杨填堰灌区有斗渠 16 条，城固县灌溉农田 6,751 亩，洋县灌溉农田 2,586 亩。

汉中古堰的工程主要以巨石为主，锁石为辅，横一大木，植以长桩。直到 14 世纪以后才以砌石为主。

汉中三堰的工程运行方式是，渠首拦河低坝将河流水位抬高，经引水口把水输入干渠，再通过分水闸或者节制闸送水至各级农渠，以灌溉农田，汛期进入渠道的洪水以及灌溉尾水，通过渠道退水闸再回归江河。

六 中国古代四大水利工程之一：它山堰

它（tuó）山堰位于浙江省宁波市鄞江镇它山旁，始建于唐代，是一座兼具阻咸、蓄淡、引灌与泄洪等多功能的区域性水利工程。它山堰与郑国渠、灵渠、都江堰合称为中国古代四大水利工程，是全国重点文物保护单位和世界灌溉工程遗产。

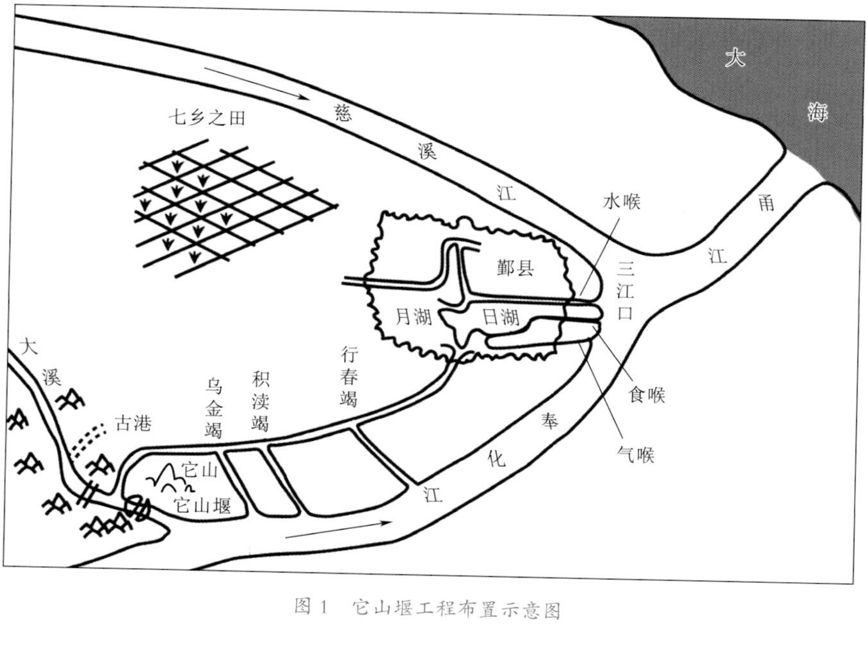

图 1　它山堰工程布置示意图

图 2　它山堰渠首

1. 功能

宁波地处沿海地区，在潮汐力的作用下，海水会倒灌进甬江及其支流鄞江之中，使得河水变为“民不能饮，禾不能灌”的咸水。唐太和七年（833 年），山东人王元暐被贬到鄞江做县令，上任后看到人们旱季无水可饮，雨季时洪涝淹没房屋和庄稼，他决心整治鄞江，组织百姓在它山下鄞江上筑堰，拒咸蓄淡，并开南塘河引水灌溉鄞西平原七乡农田，该堰因山得名“它山堰”。它山堰的建成，在非汛期，防止了咸水沿河道溯回入侵，保障鄞西平原地区的百姓和农作物不再受咸水困扰。当汛期来临时，它山堰及配套水利设施还可以将洪水从鄞江宣泄而下，保障了内河地区不受洪水侵扰。它山堰使得鄞西平原成为鱼米之乡，保障了宁波地区一千多年来的社会经济发展。

2. 堰体构造原理

它山堰建在鄞江上游出山处的四明山与它山之间，宽 100 多米，两座小山之间的山脚下的基岩可以增强堰体的抗冲击力。它山堰全长 113.7 米，堰面用条石砌筑而成，堰身为木石结构，堰面宽 4.8 米、高 10 米，堰体上部全部以长 2～3 米、宽 1.4 米、厚 0.2～0.35 米的条石砌筑而成，左右两边各有石级 36 级。据专家分析，堰身设计方面的科学性颇具现代原理，有许多原理是 20 世纪才发现的，堪称水利建筑史上的奇迹。它山堰堰体有以下四大特点。

第一，堰底向上游倾斜 5 度，多了这个倾斜角度之后，就好像堰体前面多了个小钩子，勾住了河床，可以增加堰体的抗滑稳定性一倍以上。

第二，组成堰体的条石附有黏土夹碎石层，从而减少河床的渗漏；同时，这厚厚的“古代混凝土”还可以防止涨潮时下游海水通过堰体渗透到上游去，此举一举多得，也是它山堰的神来之笔。

第三，堰体平面略向上游鼓出，从而减少了对两岸河床的冲刷。弓形水坝是现在建坝最常用的形状，国外最早的弓形水坝出

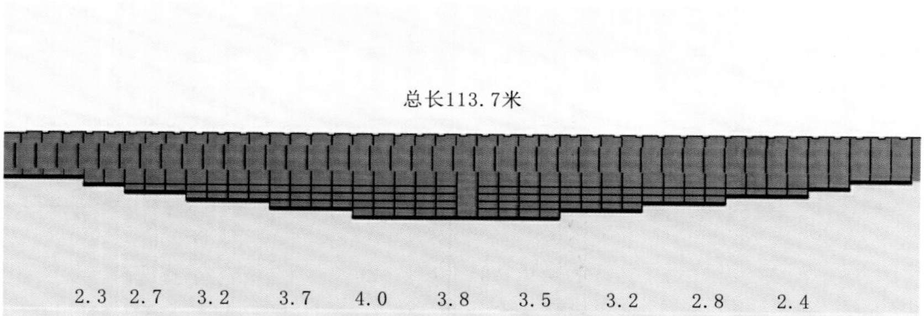

图 3　它山堰纵切面图

现在 16 世纪，它山堰比它们提前了 800 多年。

第四，堰体采用变厚布置，从而增大河床中央堰体刚度，这种布置使整条堰体沉陷均匀。

古老的它山堰，证明了古人的智慧，在那样的年代，顺自然之势而为，顺天顺地顺人心，道法自然，令今人叹为观止，为宁波的繁荣与发展做出了巨大贡献，也积淀了特有的水文化内涵。

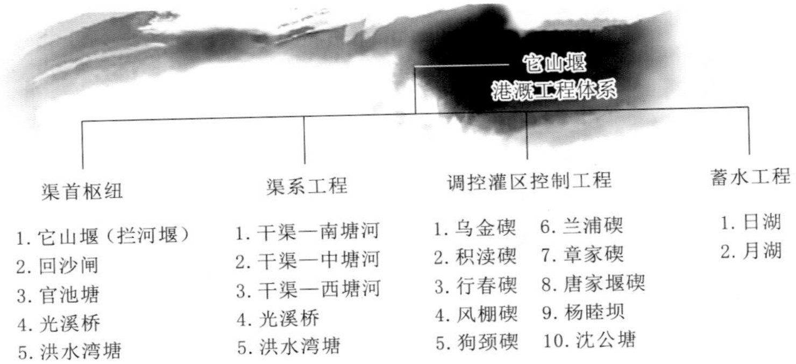

图 4　它山堰灌溉工程体系

七 我国保存最完好的古代水利工程之一：志丹苑元代水闸

志丹苑水闸建于元代，距今已有700年历史，是我国考古发现的古代水闸，也是迄今为止我国保存最完好的古代水利工程之一，位于上海市普陀区志丹路和延长西路交界处。

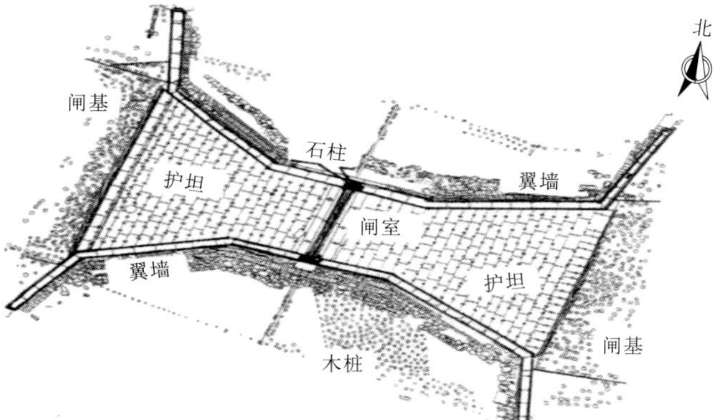

图1　志丹苑水闸遗址平面图

秦唐时期，太湖的入海水道主要是东江、松江和娄江。三江水道每日潮汐进退，将海中泥沙带入，很易淤浅。北宋时期，“今二江（东江、娄江）已绝，惟吴淞一江（松江）存焉”，太湖的入海水道只剩下吴淞江，而这一通道也逐渐淤塞。到了元代，吴淞江经历了“自然淤积—人工疏浚—河道变迁”，淤塞情况更为严峻。志丹苑水闸，这一人工疏浚工程，正是为因时因势地解决吴淞江自然淤积问题而建造的。

据考古发掘，志丹苑水闸是元代修建的、众多防止吴淞江泥沙淤积的工程之一。这些水利工程包括石闸、木闸等，主要功能是阻挡和清理潮沙及淤泥。

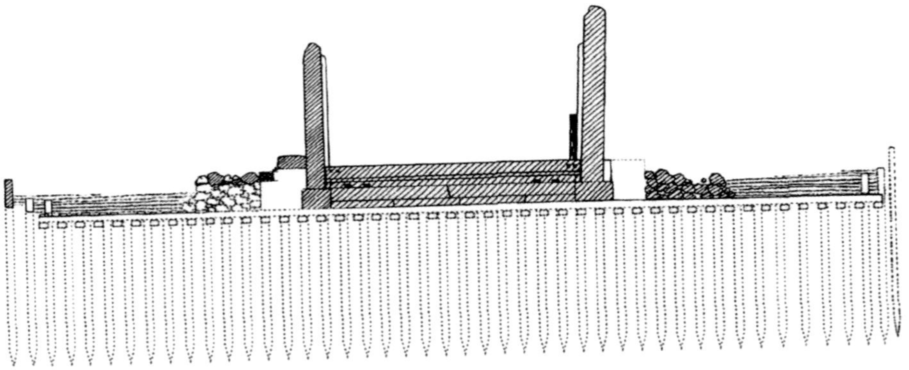

图 2　志丹苑水闸遗址剖面图

志丹苑水闸由闸门、翼墙、底石、夯土等部分组成，宽 6.8 米，总面积为 1,500 平方米，布局严谨，建造方式十分精细，用材做工俱佳。

第一步，选定建闸位置，挖出底槽后，打满一万根 4~6 米长的松木桩以加固土体，每根木桩都有编号，说明工程中每一个细节都有详细的记录。这是对工程施工过程中的一种监督措施，也反映了志丹苑水闸工程的重要性。

第二步，木桩空隙之处填嵌碎石并夯实，木桩上架木梁，铺衬石木板，木板上铺青石石板为底石，石板和石板之间用约 400

图 3　松木桩

图 4　水闸底石

只铁锭卯住。

第三步，建闸墙，立闸门石柱，闸墙基础为大石块，建在衬石木板上，其上砌多层石条，闸门石柱夹在石墙之间。

最后，在石墙外砌衬河砖并堆垒荒石，荒石外填三合土。

考古发现，志丹苑水闸的设计、施工和整体功能，以及石头与石头之间的缝隙和工艺水准，都做得精益求精。可以说，700年前的工程建造质量不输今天的水平。

志丹苑水闸对研究吴淞江、太湖流域乃至中国的水利史以及元代上海地区的经济实力，都是不可多得的实物例证，也是研究上海城镇、城市发展史的珍贵资料，在中国水利工程发展史和城市发展史上都具有极其重要的地位。

八 千年不涝古城的大功臣：福寿沟

赣州位于江西南部，章江和贡江在这里交汇，整座城市三面环水，防洪的压力非常大。福寿沟是赣州古城一个地下排水系统，俗称“鱼肠沟”，全长 12.6 公里。北宋年间（960—1127 年），赣州知州刘彝，在前人的基础上，根据赣州城的地形地势特点，采

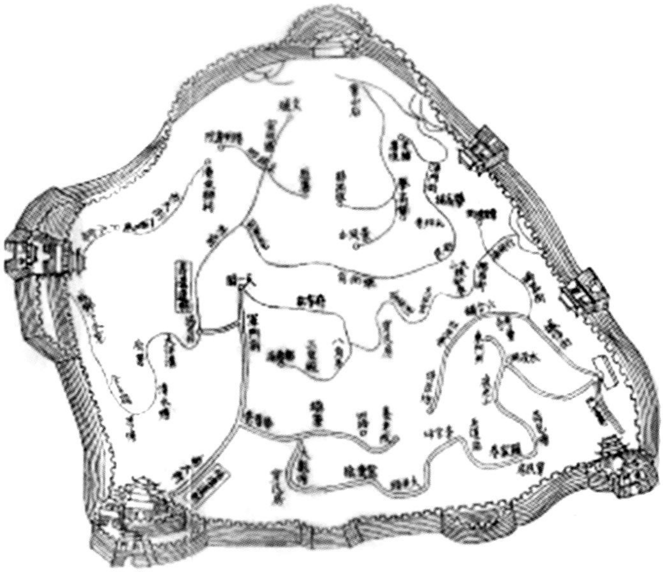

图 1　福寿沟图

取分区排水的原则，建成了两条地下沟渠，让城中的雨水和污水，顺势流入沟渠，然后由两条沟渠导引，从东西两个方向，排入城外的章、贡二江。

1077 年，福寿沟历经 10 年建成，赣州城彻底告别了水患。百姓由此保护了财产，获得了幸福长寿。同时，因为两条沟的走向形似篆体的“福”“寿”二字，当地居民形象地把这个排水系统称为“福寿沟”。“福寿沟”既是沟渠的具象，又具有良好的文化寓意。福寿沟设计建造具备以下特点：

（1）在工艺上，福寿沟选择了砖拱结构，因为砖拱桥在成本上是最低的，但引力又是最好的。利用这种结构，增加了沟墙的受力，同时又延长了使用寿命。拱形的建筑结构，和青砖、麻条

图 2　砖拱桥

石垒砌的墙体，也确保这条地下水道拥有了长久的使用寿命。

（2）师法自然，是福寿沟建筑的特点之一。赣州城地处丘陵，城内地形高低起伏，福寿沟根据地形地势，采用不同的高低、宽窄和坡度，既有利于加快水流速度，使雨水迅速排出，又有利于冲洗沟内淤积物，保证畅通。

（3）福寿沟排水系统承担了暴雨时调蓄与平时排污两大主要功能，在居民的院子与厨房里分布了各种排水设施，污水从居民家中便可直接排到福寿沟，然后流出城外。

（4）福寿沟沟底用石块铺砌，具有抵抗污水中酸、碱、热和腐蚀的抗腐效果。

（5）福寿沟地宫的铭文砖记载着工匠姓名，这是一个质量追溯体系，确保工程质量不会出问题；铭文砖虽历经近千年，字迹依然清楚，证明了古人的工艺水平和工匠精神。

（6）福寿沟的建设理念里考虑了生态效益最大化。福寿沟与城内三大池塘和几十口小塘连为一体，有调蓄、养鱼、溉圃和污

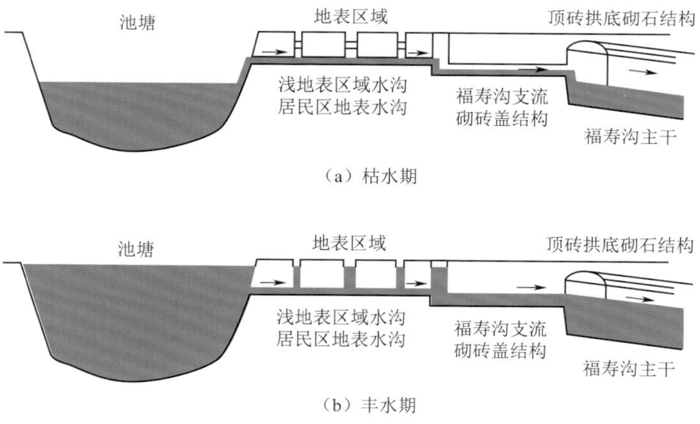

图 3　福寿沟排水示意图

水处理利用的综合功效，形成了一条生态环保循环链，和今天提出的“海绵城市”理念不谋而合。

至今，福寿沟已历时近千年，由于其形制的科学性、合理性和实用性，使得赣州据水利之便，而不受洪涝之害，千年不涝，至今还在使用。福寿沟作为中国古代城市地下排水系统的样板受到了世界的关注，同时其巧妙的水系设计对我们今天的城市规划建设仍然具有一定的借鉴价值。

九 龙游都江堰：姜席堰

姜席堰，地处浙江省龙游县灵山江下游的后田铺村，距今已有600多年，具备天人合一的理念，蕴含可持续灌溉的智慧，素有“龙游都江堰”之称。2018年，姜席堰入选世界灌溉工程遗产名录。

图1 姜席堰布局图

图 2　姜席堰现状

灵山江是龙游县境内衢江第一大支流。作为典型的山溪性河流，其河道蜿蜒曲折，两岸群山起伏跌宕，溪流暴涨暴落，难以加以利用灌溉；在汛期，江水四处蔓延至田野，给农田造成巨大破坏，严重影响了当地的生产生活。1330—1333 年，为了引水灌溉、抵御洪水，古人利用江中沙洲设计建造了两座堰。时任龙游县令的察尔可马（蒙古人）主持工程修建工作，并委托姜、席两家分别承担一座堰的建造，因此得名姜席堰。

通过充分利用江心沙洲、天然河道、河床高差等自然条件，古人将沙洲作为纽带，上接姜堰、下连席堰，最终形成引水分流、减灾灌溉、排水排沙于一体的枢纽工程。姜堰为主堰，位于灵山江的主航道，作用是抬高水位，使部分江水沿着支流进入引水渠道中，用于灌溉，如遇丰水期，超过堰体的河水直接流入灵山江下游；席堰属于支流坝堰，目的是排放过多的渠水，使其重新汇入灵山江下游。

姜席堰堰体呈方形，基础为松木卵石框架结构。在没有力学知识的年代，古人就地取材，巧妙地将松木、卵石结合，形成牢固的堰体基础，保证姜席堰屹立数百年不倒，至今仍旧发挥着重要的灌溉作用。目前姜席堰的灌区面积为 3.5 万亩，但根据历史记载，在康熙年间有 5 万余亩土地能够得到灌溉，极大地促进了当地农业、运输、商业、文化等产业的发展和兴盛。

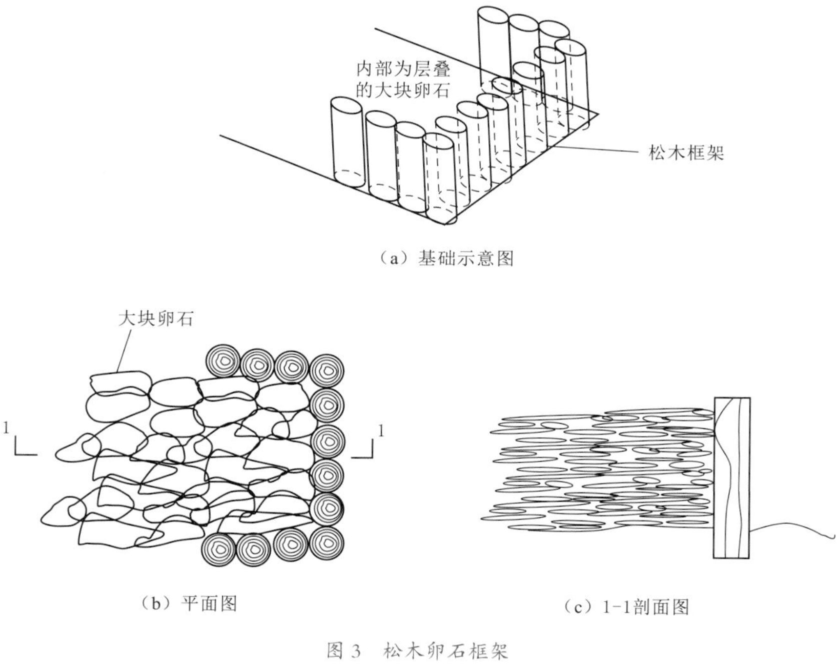

图 3　松木卵石框架

十

黄河明珠：刘家峡水利枢纽

刘家峡水利枢纽是黄河干流上的大型水利枢纽，位于中国甘肃省永靖县，为黄河上游开发规划中的第七个梯阶电站。1958 年 9 月开工兴建，1974 年竣工，是中国第一个五年计划（1953—1957 年）期间，自己设计、自己施工的大型水电工程，总投资

图 1　刘家峡水利枢纽

6.38 亿元人民币。刘家峡水利枢纽以发电为主，兼有防洪、灌溉、防凌、航运、养殖等效益，建成后成为当时中国最大的水利枢纽工程，曾被誉为“黄河明珠”。2018 年 11 月，刘家峡水利枢纽入选国家工业遗产名单。

1. 发电效益

刘家峡第一台 22.5 万千瓦机组于 1969 年 3 月投入运行，共安装 5 台机组，总装机容量 122.5 万千瓦，是中国首座百万千瓦级水电站，年发电量为 57 亿千瓦时。水电站厂房宽约 25 米，长约 180 米，有 20 层楼高，中央排列着 5 台大型国产水轮发电机组，分别担负着陕西、甘肃、青海等省供电。刘家峡水电站在中

图 2　刘家峡水电站

图 3　刘家峡水库

国西北电网中主要承担发电、调峰、调频和调压任务，是西北电网的骨干电站，处于十分重要的地位。

2. 灌溉效益

刘家峡水库总容量 57 亿立方米，控制流域面积 17.3 万平方公里，多年平均流量 834 立方米 / 秒。大坝为混凝土重力坝，坝高 147 米，坝长 204 米，顶宽 16 米。每年刘家峡水库为甘肃、宁夏、内蒙古的春灌供水 8 亿立方米，灌溉保证率由原来的 65% 提高到 85%，灌溉面积由 1,000 万亩增加到 1,600 万亩。

3. 防洪效益

刘家峡水库提高了下游梯级电站及兰州市的防洪标准，使下游盐锅峡水电站1000年一遇标准提高到2000年一遇，使兰州市100年一遇的洪峰流量从8,080立方米/秒减少为6,500立方米/秒。

4. 防凌效益

凌灾是黄河多年存在的自然灾害，每年春天解冻时，水鼓冰裂，浮冰卡坝，造成河水泛滥，堤防决口的严重冰凌灾害。刘家峡水库投入运行后，解除了下游约700公里地段的冰凌危害，近20年来没有发生重大冰凌灾害。

5. 供水效益

刘家峡水库建成后，满足了下游兰州、银川等城市的工业、城市用水需求，每天为兰州市工业供水约70万立方米。

十一

创新六项世界第一：白鹤滩水电站

白鹤滩水电站位于四川省凉山州宁南县和云南省昭通市巧家县境内，是金沙江下游干流河段梯级开发的第二个梯级电站，具有以发电为主，兼有防洪、拦沙、航运、灌溉等综合效益。

图 1　金沙江白鹤滩水电站

白鹤滩水电站是目前世界单机容量最大的水电站。水库正常蓄水位 825 米，总库容 206 亿立方米，地下厂房装有 16 台机组，总装机容量 1,600 万千瓦，年平均发电量 624.43 亿千瓦时，工程静态投资 1,430.7 亿元。2013 年主体工程开工建设，首批机组计划于 2021 年 7 月发电，2022 年 12 月工程完工。建成后，白鹤滩水电站将成为仅次于三峡水电站的中国第二大水电站。

白鹤滩水电站为金沙江下游四个水电梯级——乌东德、白鹤滩、溪洛渡、向家坝中的第二个梯级，是金沙江水电开发的骨干工程，也是继三峡水电站及溪洛渡水电站之后又一千万千瓦级巨型水电工程。白鹤滩水电站距离乌东德坝址约 182 公里，距离溪洛渡水电站约 195 公里，控制流域面积 43.03 万平方公里。

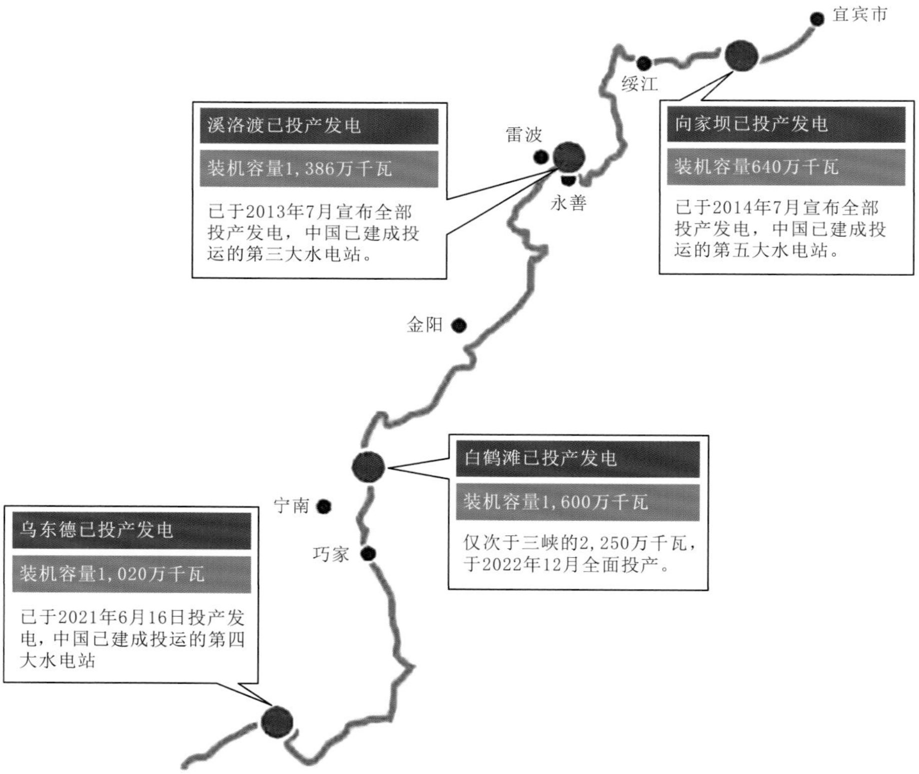

图 2　金沙江水电基地示意

图 3 建设中的水轮机组

在建设过程中，白鹤滩水电站的 6 项技术指标位列世界第一，包括水轮发电机单机容量、地下洞室群规模、圆筒式尾水调压室规模、300 米级高拱坝抗震参数、无压泄洪洞群规模，以及首次全坝使用低热水泥混凝土。

白鹤滩水电站由拦河坝、泄洪消能设施、引水发电系统等主要建筑物组成，拦河坝为混凝土双曲拱坝，坝高 289 米，拱顶厚度 14 米，混凝土浇筑量约 803 万立方米。白鹤滩水电站左右两岸分别装有我国国产的 8 台百万千瓦级水电机组，是世界上首批百万千瓦级水电机组，属于超巨型混流式水轮发电机组，可谓是世界水电行业的“珠穆朗玛峰”。据估计，白鹤滩单台百万千瓦级机组重 8,000 多吨，相当于法国埃菲尔铁塔的重量。

白鹤滩水电站建设为当地社会经济发展带来了良好的发展契机，将极大地改善基础设施建设，带动相关产业的发展；同时，工程建设对实现“西电东送”以及西部地区经济发展都具有深远的意义。

图 4　建设中的大坝

十二 金沙江上的科技明星：乌东德水电站

图 1　乌东德水电站

乌东德水电站位于云南省禄劝县和四川省会东县交界，是金沙江下游四个梯级电站（乌东德、白鹤滩、溪洛渡、向家坝）的第一梯级，也是实施“西电东送”的国家重大工程。乌东德水电站共安装12台单机容量85万千瓦水轮发电机组，总装机容量1,020万千瓦，排名世界第七，是当时世界上已投产单机容量最大的水电站，也是中国第四座、世界第七座跨入千万千瓦级行列的巨型水电站。

乌东德水电站由挡水、泄水、引水发电等主要建筑物组成。挡水建筑物为混凝土双曲拱坝，坝顶高程988米，最大坝高270米，底厚51米，厚高比为0.19，是当前世界上最薄的300米级拱坝，也是

图 2　建设中的乌东德水电站

世界首座全坝应用低热水泥混凝土浇筑的特高拱坝。乌东德水库总库容 74.08 亿立方米，调节库容 30 亿立方米，防洪库容 24.4 亿立方米。电站厂房布置于两岸山体中，各安装 6 台混流式水轮发电机组，设计年均发电量 389.1 亿千瓦时，一天发电量将满足 30 万人一年的生活用电。

2015 年 12 月 24 日，乌东德水电站全面开工。2017 年 3 月 16 日，大坝混凝土开始浇筑。2020 年 6 月 29 日，首批机组投产发电。2021 年 6 月 16 日，全部机组投产发电。目前，乌东德水电站已全面进入 945 米蓄水阶段。

金沙江下游四座水电站（乌东德、白鹤滩、溪洛渡、向家坝）总库容 458.68 亿立方米，防洪库容 154.93 亿立方米，是长江防洪体系的重要组成部分，将进一步提高川滇地区的防洪能力。

乌东德水电站是金沙江流域开发的重要梯级工程，以发电为主，兼顾防洪、航运和拦沙作用，有利于增加下游梯级电站的发电量和效益，促进当地脱贫致富和经济社会发展。

表 1　中国十大水电站

序号	水电站	流域	装机容量 / 万千瓦	年发电量 / 亿千瓦时
1	三峡	长江上游	2250	882
2	白鹤滩	金沙江	1600	624
3	溪洛渡	金沙江	1260	620
4	乌东德	金沙江	1020	389
5	向家坝	金沙江	640	307
6	糯扎渡	澜沧江	585	239
7	龙滩	红水河	490	157
8	锦屏二级	雅砻江	480	242
9	小湾	澜沧江	420	190
10	拉西瓦	黄河上游	420	102
	合　计		9165	3752

十三 实施『西电东送』的国家重大工程：溪洛渡水电站

图 1　溪洛渡水电站

图 2　溪洛渡双曲拱坝

溪洛渡水电站位于四川省雷波县和云南省永善县交界，是金沙江下游四个梯级电站（乌东德、白鹤滩、溪洛渡、向家坝）的第三梯级，具有以发电为主，兼有防洪、拦沙和改善上游航运等综合效益。溪洛渡水电站是金沙江上已建成的最大水电站，也是实施“西电东送”的国家重大工程。2016 年 9 月，溪洛渡水电站荣获“菲迪克 2016 年工程项目杰出奖”。

2007 年 4 月，溪洛渡水电站主体工程开工建设。溪洛渡水电站由挡水、泄水、引水、发电等建筑物组成，是典型的“三高三大”水电站，即高坝（285.5 米）、高地震烈度（Ⅷ度）、高速水流（接近 50 米 / 秒）、大流量（最大下泄流量约 50,000 立方米 / 秒）、大地下厂房（顶拱跨度超 30 米）和大型水轮发电机组（单机容量 77 万千瓦）。

溪洛渡水电站双曲拱坝坝高 285.5 米，是国内第三高拱坝，混凝土浇筑量约 680 万立方米。溪洛渡水电站左右岸各安装 9 台 77 万千瓦的巨型水轮发电机组，总装机容量 1,386 万千瓦，年均发电量 571.2 亿千瓦时，是目前已建成的中国第三、世界第四大水电站。

溪洛渡水库总容量 126.7 亿立方米，防洪库容 46.5 亿立方米，是长江防洪体系的重要组成部分，也是解决川江防洪问题的主要工程措施之一，使沿岸的宜宾、泸州、重庆等城市的防洪标准从 20 年一遇提高到百年一遇。

溪洛渡水电站对实现能源合理配置，优化电源结构，改善生态环境，促进西部地区特别是金沙江两岸少数民族地区的经济发展，推动长江流域经济社会可持续发展都具有深远的历史意义。

图 3　溪洛渡水库

十四 中国第五大水电站：向家坝水电站

图 1 向家坝水电站

向家坝水电站位于云南省昭通市与四川省宜宾市交界，是金沙江下游四个梯级电站（乌东德、白鹤滩、溪洛渡、向家坝）的最后一个梯级。向家坝水电站是中国第五大水电站，世界十一大水电站，也是“西电东送”的骨干电源。向家坝水电站 2006 年 11 月开工建设，2014 年 7 月全面投产发电。

向家坝水电站静态总投资约542亿元，主要由拦河坝、右岸地下厂房及左岸坝后厂房、通航建筑物和灌溉取水口等组成。拦河坝为混凝土重力坝，坝高162米，坝顶长909.26米。向家坝水库为峡谷型水库，总库容51.63亿立方米，控制流域面积45.88万平方公里，占金沙江流域面积的97%。向家坝水电站安装8台80万千瓦巨型水轮机和3台45万千瓦大型水轮机，总装机容量775万千瓦，年均发电量307.47亿千瓦时。

向家坝水电站是金沙江水电基地25座水电站中唯一修建升船机和唯一兼顾灌溉功能的水电站。此外，向家坝水电站可以解决三峡工程面临的最大问题——泥沙淤积。金沙江中游是长江主要产沙区之一，约占三峡入库沙量的1/2。向家坝水电站竣工后，三峡库区入库沙量减少了34%以上。

图2　向家坝水库

图3 升船机

在建设过程中，向家坝水电站的6项技术指标位列世界第一，包括沉井群规模、水轮发电机单机容量、单体升船机规模、洪水消力池规模、缆机跨度和砂石骨料输送带长度。

向家坝水电站以发电为主，同时兼有改善通航条件、防洪、灌溉、拦沙、反调节等综合效益，为当地社会经济发展带来了良好的发展契机，对未来改善基础设施建设、带动相关产业发展、推动长江流域经济社会可持续发展都具有深远的历史意义。

十五

万里黄河第一坝：三门峡水利枢纽

三门峡水利枢纽位于黄河中游下段，是黄河干流兴建的第一座大型水利枢纽工程，被誉为“万里黄河第一坝”。

三门峡水利枢纽1957年开工建设，1960年9月下闸蓄水。

图1 中流砥柱

图 2　三门峡水利枢纽俯瞰图

主坝全长 713.2 米，最大坝高 106 米，坝顶高程 353 米，防洪库容近 60 亿立方米，控制黄河流域面积 68.84 万平方公里，占流域面积的 91.5%。工程设计原委托苏联列宁格勒水电设计院进行，改建设计单位为天津水利水电勘测设计研究院（现为中水北方勘测设计研究有限责任公司），施工单位为水利电力部第十一工程局。

三门峡水利枢纽建成投运以来，充分发挥了防洪、防凌、调水调沙、灌溉供水和发电等综合效益，不仅是黄河下游防洪减淤工程体系的重要组成部分，也促进了当地经济社会的全面发展。因为三门峡水利枢纽的存在，黄河含沙量逐年下降；黄河下游再没有发生过凌汛决口，岁岁安澜；引黄灌溉面积呈几何式增长；

图 3　三门峡水利枢纽调水调沙

沿黄 20 个地市和 100 多个县用水无忧；黄河下游的泥沙得以排入大海。

目前，库区 200 多平方公里水域已成为国家级湿地自然保护区，对调节地区气候、保护当地生物多样性及生态环境改善起到不可或缺的作用。三门峡枢纽电站作为河南电网两个重要的大型水电站之一，装机 450 兆瓦，1973—2020 年底，累计向电网提供 560 多亿千瓦时绿色电力，相当于减少煤炭燃烧约 1,600 多万吨，减少二氧化硫排放量约 0.79 万吨，减少烟尘排放量约 4.44 万吨，为城市发展和生态环境建设做出了积极贡献。

十六　世界第八大奇迹：红旗渠

被誉为“水长城”的红旗渠是20世纪60至70年代，河南林州人民在极其艰难的条件下，沿着太行山腰修建的引漳河水到林州的水利工程。

红旗渠于1960年2月动工，30万林州人民十年如一日奋战在太行山巅，用勤劳的双手削平了1,250座山头，架设151座渡槽，开凿211个隧洞，修建各种建筑物12,408座，挖砌土石2,225万立方米，在太行山悬崖峭壁上开凿出这一人间奇迹。

图1　红旗渠

红旗渠以浊漳河为源，全长 1,500 公里，总干渠墙高 4.3 米，宽 8 米，长 70.6 公里。利用红旗渠居高临下的自然落差，兴建小型水力发电站 45 座，已成为“引、蓄、提、灌、排、电、景”相结合的大型灌区。红旗渠的建成，不仅结束了林州“十年九旱、水贵如油”的苦难历史，也从根本上改变了人们的生存条件，林州人民亲切地把红旗渠称为生命渠、幸福渠。

红旗渠的建成成为我国水利建设史上的一面旗帜。周恩来总理曾自豪地告诉国际友人：“新中国有两大奇迹，一个是南京长江大桥，一个是林州红旗渠”。红旗渠享誉海内外，被称为“世界第

图 2　建设工地

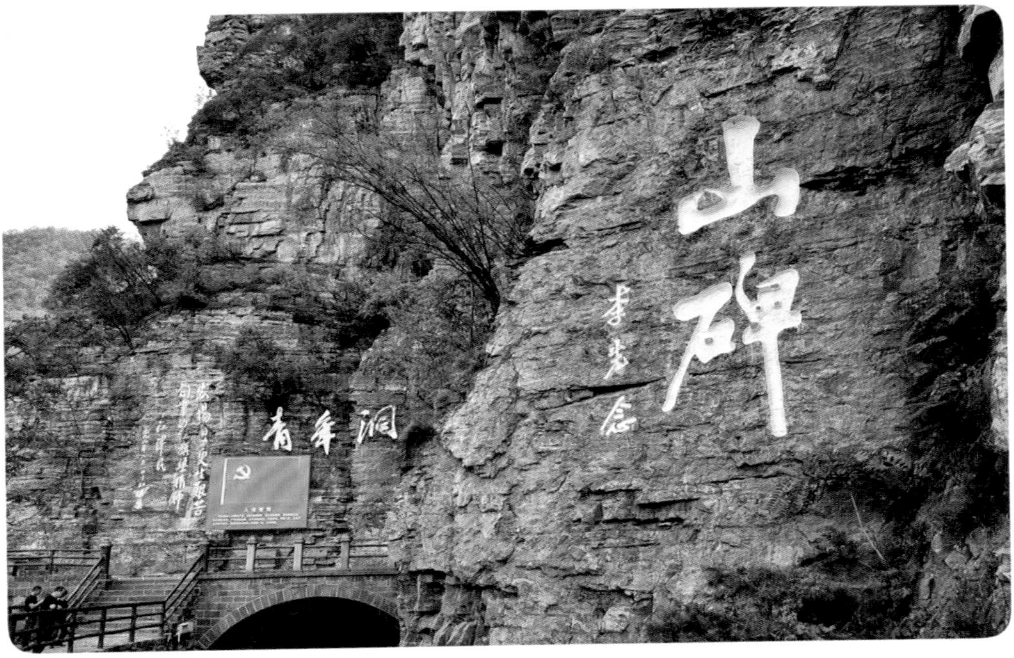

图 3　红旗渠

八大奇迹”，2006 年被国务院公布为全国重点文物保护单位。如今的红旗渠已不仅仅是一项水利工程，“自力更生、艰苦创业、团结协作、无私奉献”的红旗渠精神已升华为一种民族精神。

十七　两千余年前因战争而建的工程：长渠（白起渠）

图 1　长渠

在湖北省北部襄阳市，有一座 2,000 多年前创建的“长藤结瓜式”的蓄水引水灌溉工程，因引水干渠长约百里，称长渠。

公元前 279 年，秦国将军白起为水淹敌军，筑坝开渠。这处因战争而建的工程，很快成为襄阳平原重要的灌溉工程。为纪念它的创始人又命名“白起渠”。

图 2　秦国将军白起

3 世纪时在长渠以北修建了木渠，木渠在汉江的另一条支流鄢水上引水。长渠、木渠两渠渠系相通，在今襄阳、宜城间形成了“溉田六千余顷”的大型灌区。襄宜平原自此成为汉江中流的粮仓。

图 3　长渠与木渠相通

长渠、木渠历经秦汉至南北朝600多年的完善，至迟在5世纪时已经成为具有引水、蓄水功能的区域灌溉工程。13世纪至20世纪初以来，长渠、木渠一直沿用官方与民间自治结合的管理体系。19世纪因上游水运与下游灌溉的矛盾尖锐，渠道失修。20世纪中期，原址重建。

长渠灌溉工程体系由渠首工程、渠系工程和调蓄工程组成。渠首工程位于南漳县武安镇附近蛮河主流与支流清凉河汇合处，干渠纵贯灌区中部。

目前灌区共有15座中小型水库、2,671口堰塘。水库与长渠以沟渠相连，有闸门控制。长渠如同一条藤，沿渠与之串通的水库、堰塘，犹如一个个“瓜”，成为中国南方“长藤结瓜式”灌溉工程的典型。非灌溉季节，拦河坝使河水入渠，渠水入库、塘。灌溉季节，库塘给渠道补水，提高了灌溉的保证率。

图4 “长藤结瓜式”灌溉工程

图 5　长渠

两千年来长渠灌溉工程的持续运用，使襄阳成为天下膏腴之地，并孕育了丰厚的地域文化。在灌区的渠道两岸、陂塘湖畔分布着祭祀历代修河渠功臣的祠堂，人们以此告诫后人兴修水利、发展农耕。千百年过去，长渠仍在讲述着不老的传奇，仍在焕发着青春的活力。

十八 亚洲天池：丹江口水库

丹江口水库位于汉江（长江支流）中上游，横跨湖北、河南两省，是亚洲第一大人工淡水湖，也是南水北调中线工程水源地，有“亚洲天池”之美誉。

图 1　丹江口水库

图 2 丹江口水库大坝

丹江口水利工程始建于 1958 年，并于 2013 年完成了大坝主体加高，形成了现在的丹江口水库，总面积 864 平方公里，水库库容 290.5 亿立方米，控制流域面积 1,022.75 平方公里，正常蓄水位 170 米，多年平均入库水量为 394.8 亿立方米。丹江口水利工程由大坝、水力发电厂、升船机和两个灌溉引水渠等组成。丹江口大坝高 162 米，总长 2,494 米。丹江口发电厂装机 6 台，年发电量为 45 亿千瓦时。

丹江口水库是南水北调中线工程的重要组成部分。这一库碧水穿越黄河向河南、河北、北京、天津等四个省（直辖市）的 20 多座大中城市供水，有效缓解了中国北方水资源严重短缺的局面。

丹江口水库兼有防洪、发电、航运、灌溉等综合效益。丹江口水库缓解了武汉、襄阳等 23 个县市及 1,860 多万亩耕地的洪水威胁，改善了湖北、河南地区水上交通的条件。此外，丹江口水库灌溉耕地 360 多万亩，并为淡水鱼类养殖提供了良好条件。库区旅游业的兴盛也极大地促进了当地社会经济的发展。

图 3　丹江口水库大坝

图 4　丹江口水库

十九 “双遗产”名片：兴化垛田

兴化垛田，地处里下河平原腹地，位于中国的东部沿海江苏省，该区域河网密布，水资源充沛。早在7,000多年前的新石器时代，里下河区域还是江淮之间的一片汪洋海湾。随着海岸线的东移，该区域逐步形成了封闭的泻湖平原。

图1 兴化垛田俯瞰图

1194—1855 年，黄河侵夺淮河下游河道入海，为里下河区域带来了频繁的水患以及大量的泥沙，湖群沼泽得到淤垫。在此期间，为应对频发的洪水，当地的人们在浅水区垒土，将淤泥水草挖出堆垛，不停反复，年复一年，堆垛日渐抬高，形成了可以在上面种植的“垛田”。

垛田有一种独特的灌溉方式——戽水。高水位期间，人们可以行船直接为两岸垛田戽水浇灌，低水位期间，人们通过梯级戽水的方式为垛田灌溉，最高的垛田有 4～5 级。垛田水位夏季秋季高，冬季春季低。人们在冬季春季垛田水位较低的时候，通过罱泥等方式用垛田间河底的淤泥来堆叠垛田，以防备夏季秋季可能发生的洪水灾害。

图 2　农民在兴化垛田中劳作

图 3 兴化垛田

在夏季秋季，人们把垛田中收获作物后的其他部分投入水中，沤制成为河底富含有机质的淤泥。到了冬春季，罱泥也具备了施肥的效用。充沛的水量加之高肥力的土壤，垛田自古就盛产优质的瓜果蔬菜，现在垛田生产的蔬菜已成为当地农民经济收入的主要来源之一。

垛田地区独特的水利景观、良好的生态环境和多彩的民俗文化是旅游产业的宝贵资源。菜花垛田、水上森林、湿地公园以及渔业生态园等著名景区，也为当地经济发展提供重要助力。2022 年 10 月，兴化垛田被列入世界灌溉工程遗产名录。2014 年 4 月，兴化垛田被联合国粮农组织正式评定为“全球重要农业文化遗产”。

二十 最大的客家梯田：上堡梯田

上堡梯田位于江西省赣州市崇义县西北部山区，面积约有3,400公顷，主要分布在上堡、丰州、思顺三个乡（镇）26个行政村。梯田最高海拔1,260米，最低280米，垂直落差近千米，

图1 上堡梯田俯瞰图

图 2 上堡梯田

最高达 62 梯层，2012 年，被上海大世界基尼斯评为“最大的客家梯田”，2018 年 4 月，被联合国粮农组织列为“全球重要农业文化遗产”。

据《山海经》等文献记载，上堡梯田开发历史最早可追溯至先秦时期，梯田兴于秦汉时期，距今至少 2200 年历史，后经唐、宋、元、明、清代的不断扩建和修缮，至此达到现在宏伟的规模。上堡乡赤水村至今保留的石狗遗存证明了梯田的历史。

“上堡上堡，高山梯田水淼淼。”上堡梯田因山成形、因水而兴。高山地带厚密的森林植被，涵养了丰富的水源；梯田区的花岗岩山体构造，构成了一个不透水的蓄水之库，山顶和山坡所截留的雨水，只能从山腰坡地渗出，形成了天然的地下输排水网络，为梯田提供不竭的水源。

上堡梯田多为自流灌溉，一般根据孤立山头或者季节、气象变化输水需要，采取不同的灌溉方式。据现存遗迹，上堡梯田灌溉系统主要分为三大类 12 种方式，其中蓄水工程 3 种（高山流水、蓄水分流、过滤沉砂），灌排渠系 7 种（筑坝引水、逆田走水、借田输水、之型流灌、架桥输水、竹筒出水、水车提水），水量调节控制设施 2 种（分水口石块控制、管道闸阀控制）。

时至今日，上堡梯田自流灌溉体系的工程形式和布局仍然完整的保存，这些凝聚着先民汗水和智慧的典型梯田灌溉系统，成为当地农业经济的重要组成部分，并在现代农业发展理念的结合下，推动上堡梯田发挥其独特的价值功能。

二十一 活态博物馆：松古灌区

松古灌区位于浙江省松阳县，是瓯江流域最大产粮区，总灌溉面积 16.6 万亩。古谚“松阳熟，处州足”，是对松古灌区物阜民丰的写照。

图 1　松古灌区工程之一——午羊堰

公元前138年，东瓯国北迁，驻扎松阳县古市镇。因当地灌溉条件优越，部分军民开垦耕作，古市镇自此农耕发达、商贾繁荣。松阳农业灌溉史由此开篇。1041—1048年间，松阴溪南岸、独山脚下已建成百仞堰；十三都村先后修筑芳溪一堰、二堰，灌溉八千余亩农田；1340年，松阳城南建成“白龙堰”，灌田两千余亩。宋元时期，松古灌区初具规模。

图2　松古灌区

明末清初，松阳县境内有古堰116多处，居瓯江流域各县之首。松阳先民依势筑堰建渠，逐步建成以松阴溪为水源的松古灌区，拥有完备的管理制度和堰堤密布、圳渠交错的灌溉网络。目前，松古灌区内灌溉千亩以上的古堰，现存14处，数量之多，实属罕见。

珍藏在松阳水利博物馆的14位县令的16帧告示、榜文和碑记，记录了555年间松古灌区修复、管理、分水、水纠纷等史实，也记录了数百年来当地官民执着追求的治水精神。

二十二 千年大运河的最北端：通惠河

通惠河属京杭大运河北运河水系，亦名通济河，元时称金水河，明代以后改称御河。通惠河全长约 82 公里，东西走向，始建于元代（1206—1368 年），是由郭守敬主持修建的漕运河道，元世祖将此河命名为通惠河。

图 1　清末的通惠河

图 2　旧时通惠河通州河口段

通惠河的全盛时期并不很长，随着元朝的灭亡，明初建都南京，通惠河一度曾被废弃。到明永乐年间（1403—1424 年）迁都北京，才又开始对通惠河进行治理。明代曾对通惠河进行了几次较大规模的治理和改造，除漕运船只不能直接进入北京外，其他基本恢复了元朝时期漕运的盛况，一直沿用到 20 世纪初叶。

漕运停止后，通惠河承担着农业用水的输水任务，也曾是北京城区排水面积最大的水系。随着工业化的发展，通惠河成为北京城区污水排放的主河道，后来污水被排放到高碑店污水处理厂，通惠河水质才逐年改善。经几十年的治理，通惠河又成为了北京市的景观河道。

图 3　现代的通惠河

图 4　现代的通惠河

目前通惠河全长21公里，穿过北京东二环到东五环的核心区域，连通北京站、CBD、文创试验区等核心功能区。其流域面积为250平方公里，河道最大排水能力为350立方米/秒，主要支流有护城河、城内河系、金河、长河及南旱河。

通惠河曾是北京的一条经济命脉，对繁荣首都经济发挥了重要作用。如今，通惠河是连接首都与通州行政副中心的水上重要通道。2020年10月，通惠河高碑店湖段旅游通航。通惠河的通航对于改善沿岸生态环境，打造京通运河旅游带具有重要意义。

二十三 中国古代第一坝：戴村坝

戴村坝位于山东省东平县境内大汶河与大清河的交汇处，被誉为“中国古代第一坝”，与“都江堰”齐名（都江堰是世界上最古老的无坝引水工程）。它始建于明永乐九年（1411 年），至今已有 600 多年的历史。戴村坝的主要功能是引汶济运，确保大运河

图 1 戴村坝俯瞰图

图 2　戴村坝

南北贯通，对明清两朝的政治统一、经济发展和文化交融发挥了不可替代的作用，其功能等同于人的“心脏”，故被誉为“运河之心”。2014 年，戴村坝在第 38 届世界遗产大会被列入《世界遗产名录》。

明永乐九年（1411 年），明成祖朱棣诏令工部尚书宋礼督工，重开运河。宋礼征调民夫 16.5 万余人，新开河道 120 里。起初，宋礼等官员无法处理运河时常干涸的问题。最后，他们采纳了民间治水专家白英的建议，筑戴村坝以遏汶水入海之路。大坝修成后，拦汶水使其顺小汶河南下，再分水南北。从此，妥善地解决了丘陵地段运河断流的现象，帮助大运河完成其交通大动脉的使命。此外，在夏秋水丰之际，戴村坝可以漫入大清河来疏溢；反之在春冬水涸之际，它拦截南流河水，防止大汶河干涸。

戴村坝全长1,599.5米，控制流域面积8,537平方公里，约占大汶河流域总面积的80%。戴村坝由主石坝、窦公堤和灰土坝组成，三者互为利用，互相保护。主石坝呈阶梯状，由玲珑坝、乱石坝和滚水坝构成。主石坝为石结构，使用重达数吨的巨石，石与石之间采用束腰扣结合法，一个个铁扣把大坝锁为一体。

历经数百年，任洪水千磨万击，在今天戴村坝仍铁扣紧锁，岿然不动，这充分表现了以白英为代表的我国劳动人民在治水领域的无穷智慧。19世纪初，荷兰水利专家方维在参观戴村坝后高度称赞它是十四、十五世纪的伟大工程。1965年11月，毛泽东主席称赞戴村坝是一个了不起的工程，并赞赏当年策划、主持修建这一工程的白英为“农民水利家”。在功能作用、设计思想、建筑构造和施工工艺领域，戴村坝具有极高的研究价值，是我国水利建设史上的一大奇迹。

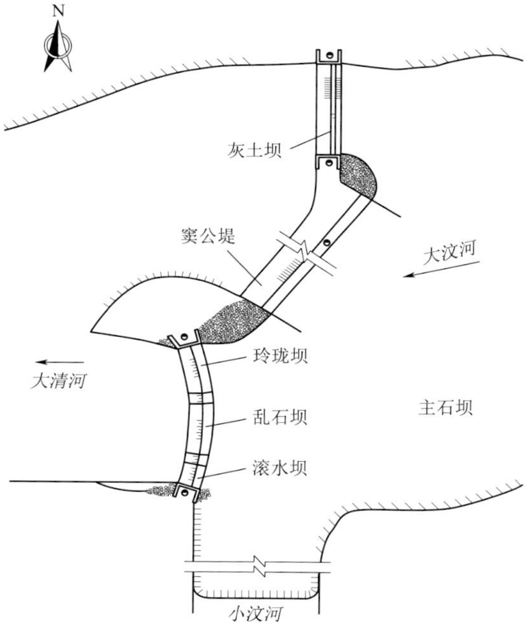

图3　戴村坝示意图

Series of China's Achievements in Water Projects

Water Projects in China from Ancient Times II

International Economic & Technical Cooperation
and Exchange Center, Ministry of Water Resources

中国水利水电出版社
China Water & Power Press
· BeiJing ·

The Compilation Committee of Water Projects in China from Ancient Times II

Chairman: HAO Zhao

Vice Chairmen: XU Jing CHI Xinyang

Members: HOU Xiaohu LIU Bin SHEN Kejun CHANG Yuan

XIONG Jia ZHANG Linruo WANG Jinling LIU Yuwei

FOREWORD

Water is the source of life, the essence of production and the foundation of ecology. At all times and globally, achieving the benefits of water development and mitigating the impacts of water disasters have all along been a top priority of governing a country. In a certain sense, the way of governing water mirrors the art of a country's governance. The five-thousand-year civilization of China has witnessed how water governance has safeguarded people's livelihood, shaped its culture and promoted its prosperity. For China, with an indomitable history of water governance, water defines its national development.

For thousands of years, the magnificent water projects have seen Chinese people's endeavor in defending floods and droughts and developing water resources, such as Karez—the largest and most complex underground irrigation project, Danjiangkou Reservoir—renowned as "the Asia's Heavenly Lake", Huang Ju Irrigation Project—a World Irrigation Project Heritage, the Red Flag Canal—"the eighth wonder of the world", among others.

As this book unfolds, you will have a better understanding of China's journey of water governance and its water culture, and the significance of water resources as the lifeline of agricultural, economic and social development. More importantly, you will come to realize that water security plays a pivotal role in promoting the nation's long-term stability and rejuvenation.

Zhao Hao

November 2023

CATALOGUE

I

The Karez — the world's largest and most complicated underground irrigation project

The Karez, invented in the Western Han Dynasty, has used as underground irrigation system for more than 2,000 years. It is a great invention of the working people in China to adapt to dry climate with high evaporation and low rainfall.

The main design principle of the Karez is to transport water through an artificial underground channel from the aquifer to the above-ground channel for irrigation. The project is composed of four parts: vertical

Fig.1 The entrance of the Karez

access shafts, an underground channel, an above-ground channel and a pond. Water in the aquifer mainly comes from melted snow that infiltrates into the ground. People would first find an aquifer and design a route to distribute water. Then a number of access shafts shall be built and connected by underground channels. The water flow makes full use of the natural slope or gravity to reach the irrigation area. This kind of irrigation system reduces the evaporation loss along the way to the greatest extent. At the time of no guide systems and only simple laboring tools being applied, it is indeed a great pioneering work that accomplished the goal totally relied on manual for underground operation.

Today, in Xinjiang region of China, more than 1,700 Karez are still functioning, with an irrigation area of about 500,000 mu, most of which are concentrated in Turpan Area.

Karez is known as one of the three marvelous engineering of ancient China with no parallel except the Great Wall and the Grand Canal.

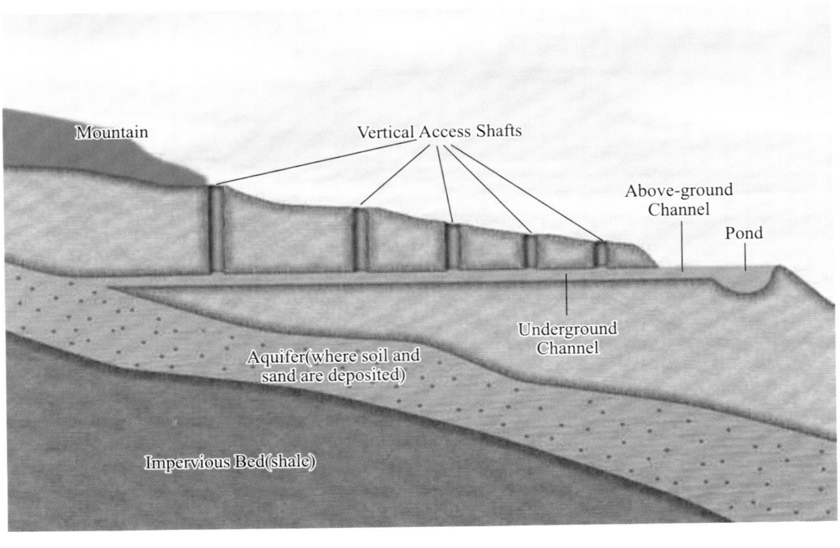

Fig.2 Structure of the Karez

II

Tongji Weir—the world's oldest arch dam

Initiated in the 4th year of the reign title of Tianjian (A. D. 505) of the Liang Dynasty (Known as the Southern Liang, A. D. 502—557) and rebuilt in the 1st year of the reign title of Kaixi (A. D. 1205) of the Southern Song Dynasty (A. D. 1127—1279), Tongji Weir, situated in Songyin Stream, upstream of a tributary of Oujiang River in the southwest of Zhejiang Province, is the world's oldest arched water project mainly serving irrigation.

At first, Tongji Weir was built of wood. In Song Dynasty He Dan, an official who resigned from office and went back to his hometown, requested the Emperor to rebuild Tongji Weir, and the rebuilt Tongji Weir was changed to stone dam. The dam is 275 m long, 25 m wide and 2.5 m high. The upstream rainwater collection area is about 3,150 km^2, the water diversion flow is 3 m^3/s, and the whole dam is convex to the upstream about 120 degree arc. The arch dam can better withstand the flood force. Its construction time is earlier than that of Elche arch dam built in the 16th century in Spain and Ponte Alto arch dam built in 1612 in Italy.

In addition to the arch dam, this water project boasts of many characteristics. The first is to fully consider the natural elevation difference when selecting the dam site, so that the supporting irrigation channels can irrigate by gravity. The second is to use big pine as the dam foundation and cast stone dam with molten iron to ensure that the

foundation is not rotten and the overall performance of the stone dam is strong. This is also one of the important reasons why the dam lasts for thousands of years. The third is to give consideration to both sand discharge and navigation. At the north end of the dam, there are two sand discharging gates with a clear width of 2 m, which are deep to the bottom of the dam. The sand and stone washed down by the upstream flood will be automatically discharged to the downstream of the dam by rapid flows of these two gates. A ship lock with a clear width of 5 m was built to ensure the navigation function in ancient times. Fourth, an overpass stone bridge for water diversion was built, commonly known as the "Three Hole Bridge", to facilitate the mountain stream flood affecting the safety of the irrigation channels to pass through the bridge

Fig.1 A bird's eye view of Tongji Weir

Fig.2 The water-channeling bridge of the Tongji Weir

deck into the Oujiang River, and the irrigation water to flow under the bridge. Therefore, there is no mistake in flood diversion and irrigation. Fifthly, the irrigation water volume can be adjusted reasonably through 72 gates and 3 connected canals. From the Song and Yuan Dynasties to the Qing Dynasty, Tongji Weir was maintained and renovated many times.

The historical records and rules of Tongji Weir have been well preserved. Since its launch, all dynasties had attached great importance to its maintenance and management with a set of self-contained and complete management methods. Among them, the earliest written "weir rule" appeared in the 7th year of the reign title of Yuanyou (A. D. 1092) of the Northern Song Dynasty (A. D. 960—1127). The earliest existing

weir rule was formulated by Fan Chengda, governor of Chuzhou, in the 4th year of the reign title of Qiandao (A. D. 1168) of the Southern Song Dynasty, and was unique, scientific and comprehensive. It was imitated by later generations. Now it is preserved in the Zhannan Sima Temple beside the dam. The rules cover the standard sizes of branch outlets, rotation irrigation system, annual maintenance and labor service, methods for apportioning labor and materials, methods for rewards and punishments in project inspection, etc. Until now, some rules are still in use.

Tongji Weir is now surrounded by breathtaking sceneries. The Yantou Village where the river originates is cradled in mountains and covered by canopies of ancient camphor trees. The Wenchang Pavillion stands next to the "Three Hole Bridge". To the west of the dam is the Zhannan Sima Temple, more famously known as the Dragon Temple. It keeps 16 stone tablets of the Song, Yuan, Ming and Qing Dynasties and the Republic of China, recording the construction history as well as rules and maps of the dam. Tang Xianzu (a dramatist and litterateur of the Ming Dynasty in China) and other famous Chinese scholars and calligraphers left their works here.

III

Huang Ju Irrigation Project — constructed for hundreds of years, and irrigated thousands of farmlands

Huang Ju Irrigation Project lies in Huotong Town, Jiaocheng District, Ningde City, Fujian Province. Over 1,400 years ago, the Huang family, led by Huang Ju, an official in charge of admonition in Sui Dynasty (A.D. 581—618), built it along the Huotong Stream. It is the most complete water project with the highest technical level in Sui Dynasty. In fact, besides its irrigation function, the Project also plays a role in domestic water supply and makes full use of potential water

Fig.1 The Huang Ju Irrigation Project

Fig.2 The Pipa Tunnel

energy, so it is also named as a small comprehensive water project with multiple purposes. In 2017, the Project was added to the World Irrigation Project Heritage List.

The Project consists of the Pipa Tunnel water diversion project on the left bank and the Longyao Canal irrigation project on the right bank, extending over 10 km and watering an area over 20,000 mu (1,333 hm^2). The Pipa Tunnel is over 700 m long in total, 2.41 m high and about 1 m wide in average. Sediment discharge holes were drilled along the Huotong Stream to facilitate dredging and sediment discharging.

The trunk canal of the Longyao Canal is an open channel with a total length of over 5,000 m, a width of 1.51–2.72 m and a depth of 0.95–3.00 m.

Fig.3 The Longyao Canal

Five-stage water hammers were built by the means of using water level difference, and water wheels drove grinding, crushing, pounding and sieving tools to process agricultural byproducts, grains and oil. What's more, this waterway was then divided into two branches. One was used to irrigate the farmlands of over 1,000 mu (67 hm^2) in Shiqiaoyang, while the other was employed to provide domestic water to local villagers, forming a complete system of irrigation, waterpower and water supply for people's livelihood.

The technical accomplishment of the Project is mainly reflected in the following. Despite the harsh geographical condition in an unpowered driven era, the ancient people, by taking advantage of elevation difference and through scientific planning, dammed water and diverted water through open channel and drilled tunnel, forming the rational layout of diversion, transfer, storage, irrigation and drainage and realizing the multiple functions of this irrigation project. The rock-cutting method of baking by fire and freezing with water was adopted, thus shaping the proper cross section of tunnels and stepping up the structure stability.

The Project is a model project built by local community with self-raised funds and guided by the government. Prospering for over 1,000 years, it still retains its original appearance and plays an integrated function in agricultural irrigation, domestic water supply and water power processing, thus becoming the historical witness of irrigation civilization for present and future generations.

IV

Cross-Yellow River Project of the South-to-North Water Diversion Project (the Middle Route) — linking the Yangtze River and the Yellow River

Located in Gubozui, about 30 km west of Zhengzhou City, Henan Province, the Cross-Yellow River Project is a landmark in the middle route of the South-to-North Water Diversion Project, which refers to the underground tunnel across the Yellow River. The middle route of the South-to-North Water Diversion Project is supposed to channel the Yangtze River from the Danjiangkou Reservoir all the way to the north, but the Yellow River forms a natural barrier to block the way of the south water to the north. To tackle this problem, the Cross-Yellow

Fig.1 Scheme of the Cross-Yellow River Project

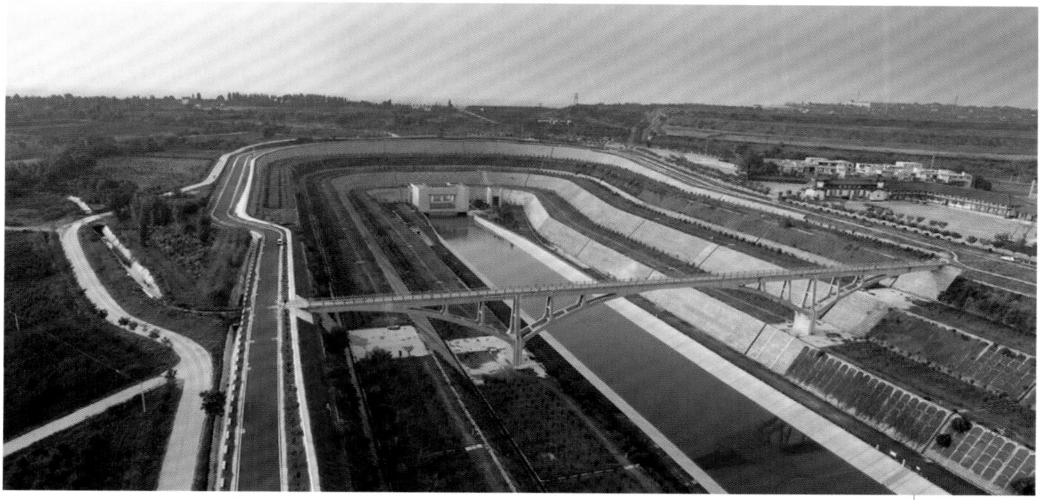

Fig.2 The water inlet (the south bank)

River Project, the largest river-crossing water project in human history, was built up.

The project has a total length of 19.3 km, including a 13.95 km open channel, a 4.71 km tunnel and 0.65 km buildings. The "tunnel under the river" mode is employed in the crossing-river section. Two tunnels with a diameter of 7 m each are set in parallel at a depth of 40 m under the river bed to cross the Yellow River. The double tunnels can ensure flow increase when required and continuous water supply in case of accidents. By the principle of "inverted siphon", the water inlet (the south bank) is higher than the water outlet (the north bank), and thus water will flow towards the outlet in the wake of pressure disparity between the two ends. The entrance is an inclined shaft and the exit is a vertical one, facilitating sound operation and maintenance, and saving investments.

The tunnels withstand both the external water and soil pressure and the internal flowing water pressure, so its wall adopts two layers of reinforced concrete lining, respectively bearing external and internal

pressures. The outer layer is 0.4 m thick assembled segment structure lining and the inner is 0.45 m thick reinforced concrete prestressed one. The two layers of lining are separated by permeable cushion. A large number of monitoring devices and instruments are installed in the tunnels to carry out operation monitoring of the tunnel in real time. The anti-seismic design capacity of the Cross-Yellow River project is grade 8, and its flood control capacity is designed according to the 300-year flood and checked according to the 1000-year flood of the Yellow River.

The tunnels are excavated by Tunnel Boring Machine (TBM), with a diameter of 9 m, a length of over 80 m and a total weight of 1,100 t. In the process of construction, a series of technical problems are addressed, such as unstable river bed of the Yellow River, complex geological conditions and sand liquefaction during vibration. It is the first time that the slurry balanced pressure shield is put into place for tunneling, and many new cutting-edge technologies, including double lining structure, are practiced.

Fig.3 The water outlet (the north bank)

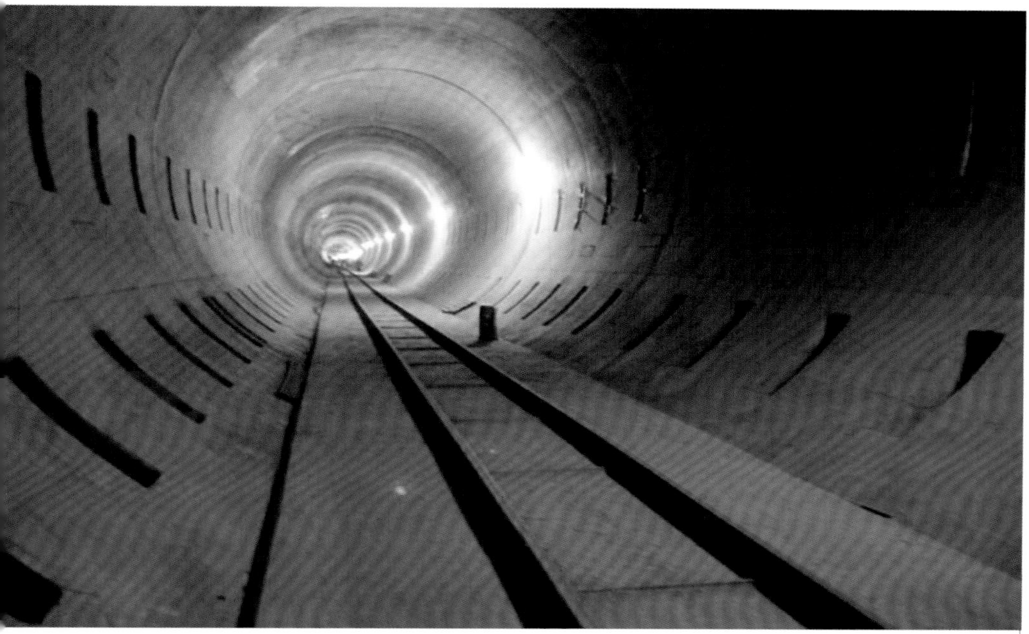

Fig.4 Inside of the tunnel

Many block stones and ancient tree trunks in the river bed, and changes in the stratum composition and water pressure all pose threats to the safety of construction.Therefore, during the tunnel excavation, the TBM parameter, thrust, torque, speed and penetration angle are constantly adjusted in accordance with the variation of geological conditions.

The Cross-Yellow River Project realizes the interchange of water from the Han River and the Yellow River, presents the magnificent landscape of river meeting, and embodies the wisdom and courage of the Chinese nation. The middle route of the South-to-North Water Diversion Project has realized continuous and safe water supply for over 1,900 days since its official operation, cumulatively diverting 30 billion cubic meters of water and directly benefiting more than 120 million people. It becomes the new lifeline of water supply in many cities with remarkable economic, social and ecological benefits.

V

Three Weirs of Hanzhong — the earliest farmland irrigation systems in Hanzhong

Lying in the Hanzhong Basin, Hanzhong City of Shaanxi Province, the Three Weirs, namely Shanhe, Wumen and Yangtian, have functioned as important irrigation schemes. Built in the Western Han Dynasty (206 B. C.—A. D. 25), more than 2,000 years ago, and as the earliest farmland irrigation systems in the region, they not only provide water for farmland irrigation in the northern part of the Hanjiang River, but also turn the Hanzhong Basin into one of the earliest "land of abundance". The Three Weirs of Hanzhong have been continuously renovated and maintained, and still irrigate 217,500 mu (14,486 hm^2) of farmland today. In 2017, the Three Weirs of Hanzhong were added to the World Irrigation Project Heritage List.

Fig.1 The canal head of the Wumen Weir

The Shanhe Weir, the first of the Three Weirs, is seated in the Baohe River, the primary tributary of the Hanjiang River that is a tributary of the Yangtze River. It is said that Xiao He and Cao Can, famous generals of the Western Han Dynasty, presided over its construction. The Shanhe Weir possesses three canal heads to intercept the Baohe River for farmland irrigation. The modern irrigation system, the Baohui canal, was basically built along the original route of the Shanhe Weir. Since the completion of the Shimen reservoir in 1975, all the farmland once irrigated by the Shanhe Weir has been covered by the Shimen South Canal.

The Wumen Weir, the second of the Three Weirs, is situated on the right bank of the Xushui River, a tributary of the upper Hanjiang River, 15 km north of the Chenggu County, Shaanxi Province. The name of Wumen means five water gates placed horizontally at the head of the canal. The construction of the weir was also started in the Western Han Dynasty. In its initial phase, the irrigation area was 3,000 mu (199 hm^2) more or less. After constant renovations and expansions in previous dynasties, the irrigation area exceeded more than 50,000 mu (3,330 hm^2) and had certain flood control capacity in the Qing Dynasty (A. D. 1616—1911). Up to now, it still performs the irrigation function.

The Yangtian Weir, the third of the Three Weirs, lies in the left bank of the middle reaches of the Xushui River, 10 km north of the Chenggu County, Shaanxi Province. It was called the Zhangliang Canal before the Northern Song Dynasty (A. D. 960—1127). From 1163 to 1165 in the Southern Song Dynasty (A. D. 1127—1279), Yang Congyi dredged the channel again, so it was renamed as the Yangtian Weir. At that time, it was able to irrigate 7,000 mu (466 hm^2) in Chenggu County and 18,000 mu (1,200 hm^2) in Yang County. Later, the Weir was destroyed by floods and rebuilt many times. At present, the

Fig.2 The water gate of the Yangtian Weir

irrigation scheme of the Yangtian Weir has 16 lateral canals, watering 6,751 mu (450 hm^2) of farmland in Chenggu County and 2,586 mu (172 hm^2) in Yang County.

The materials of ancient weirs in Hanzhong mainly rely on boulders that are supported by locked stones, with a large plank of wood placed horizontally and long piles planted. It was not until the 14th century that masonry began to dominate in the sphere of weir building.

The operation pattern of the Hanzhong Three Weirs is that the low dam at the canal head raises water level of the river, so that water is channeled into the main canal through water intakes and consequently enters into agricultural canals at all levels through water diversion gates or regulating sluices for watering farmland. Flood water entering the canal during the flood season and the tail water of irrigation may return to the river through the escape sluices.

VI

Tuoshan Weir — one of the four most important water projects of the ancient China

Tuoshan Weir is located by Tuoshan (Tuo Mountain) in Yinjiang (Yin River) Township of Ningbo City in Zhejiang Province. Initially built during the Tang Dynasty, it is a regional water project with multiple functions including interception of salt water, storage of fresh water, water diversion for irrigation and discharge of floods. Tuoshan Weir, Zhengguo Canal (246 B.C., Shaanxi Province), Lingqu Canal (219 B.C., Guangxi Zhuang Autonomous Region) and Dujiangyan (251 B.C., Sichuan Province) are collectively referred to as the four

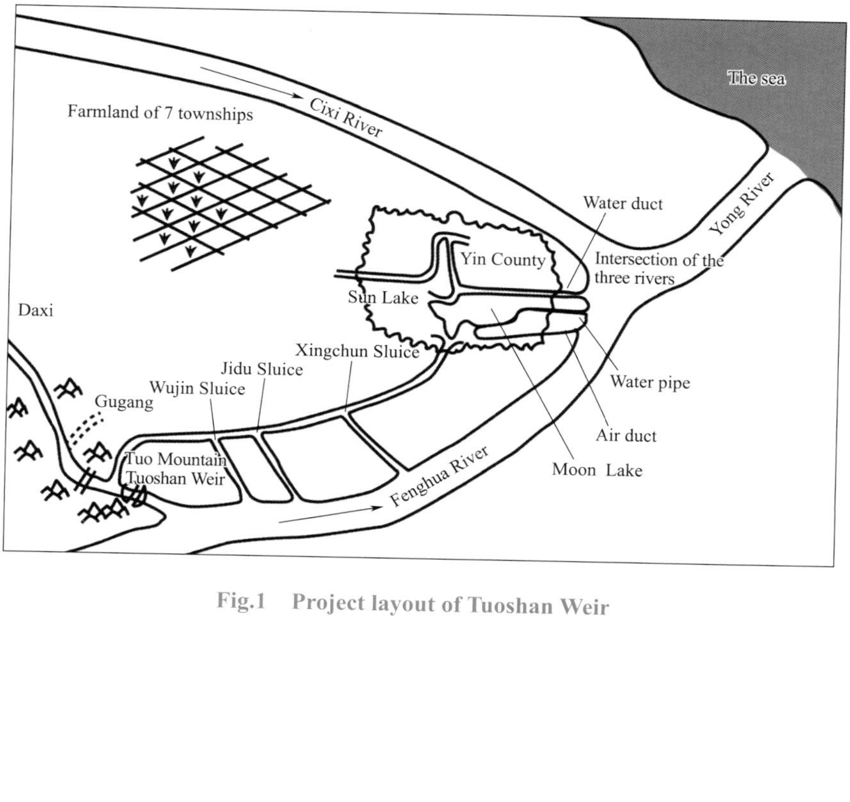

Fig.1 Project layout of Tuoshan Weir

most important water projects of the ancient China. Tuoshan Weir is inscribed as a major historical and cultural site protected at the national level and a world heritage irrigation project.

1. Functions

Ningbo is a coastal city where seawater backflows into the Yong River and its tributary Yin River under the effect of tidal forces, turning the river flows into salt water which was "suitable for neither human drinking nor crop irrigation". In the 7th year of the reign title of Taihe (A. D. 833) of the Tang Dynasty (A. D. 618—907), Wang Yuanwei, a Shandong native, was demoted to serve as county magistrate of Yinjiang. After taking office, upon witnessing people with no water to drink during the dry seasons and houses and crops flooded during the wet seasons, Wang was determined to train the Yin River. He organized the local people to build a weir at the foot of the Tuo Mountain on the Yin River for interception of salt water and storage of fresh water, and excavated the Nantang River to divert water for irrigation of the farmland in the seven townships along the Yinxi Plain (the plain area west to the Yin River). The weir, named after the mountain, came to be known as Tuoshan Weir. Once completed, Tuoshan Weir prevented salt water from backward infiltration along the river course and thereby guarded the people and crops in the Yinxi Plain against the disruption of salt water during non-flood season, and discharged flood down from the Yin River together with its supporting facilities to protect the inner river areas from flood hazards during the flood season. As a result, Tuoshan Weir made the Yinxi Plain a land of fish and rice, and has safeguarded the social and economic development of the greater Ningbo area for more than one thousand years.

Fig.2 The weir head of Tuoshan Weir

2. Structure Characteristics

Tuoshan Weir, built between the Siming Mountain and the Tuo Mountain at the outflow point of the upper reach Yin River and with a width of more than 100 m, has the bedrock beneath the mountain foot between the two hills to strengthen the impact resistivity of the weir structure. Tuoshan Weir stretches for a total length of 113.7 m, with a masonry top made of stone stripes and an internal structure that is a mixture of wood and stone. The weir top is 4.8-meter wide and 10-meter high. The upper part of the weir body is masonry made exclusively of stone stripes that are 2−3 m long, 1.4 m wide and 0.2−0.35 m thick, with both its left and right sides featuring 36 stone steps. According to expert analysis, the weir design is scientific and reflects modern principles that were mostly discovered in the 20th century. As such, Tuoshan Weir amounts to a miracle in the history of hydraulic architecture. The weir structure exhibits four characteristics.

First, the bottom of the weir tilts 5 degrees to the upstream. This tilting angle acts like a small hook in front of the weir structure, hooking the riverbed and doubling the anti-sliding stability of the weir.

Second, a layer of clay and gravel is affixed to the stone stripes of the weir structure, reducing seepage of the riverbed. At the same time, this thick layer of "ancient concrete" also prevents downstream sea water from infiltrating into the upstream through the weir structure in times of high tides. This approach, killing several birds with one stone, is a magic touch in the construction of Tuoshan Weir.

Third, the surface of the weir protrudes slightly towards the upstream, reducing erosion of the riverbed on both banks. With its dam in the shape of a bow, the most commonly used shape for contemporary dam construction, Tuoshan Weir was ahead of its earliest overseas counterpart (which came into being in the 16th century) by more than 800 years.

Fourth, thickening arrangements in the weir structure increase the rigidity of the weir body in the middle of the riverbed and even out subsidence of the entire weir.

The ancient Tuoshan Weir is a proof of ancient human wisdom. Back then, humans lived and worked in harmony with nature, following the natural rules to meet human needs and inspiring awe from generations to

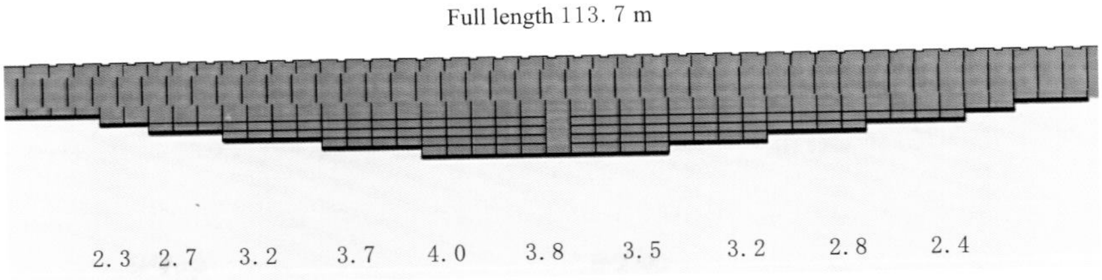

Fig.3 Longitudinal section of Tuoshan Weir

come. Tuoshan Weir leaves a mark in history with its great contributions to the prosperity and development of Ningbo and its accumulative implications of a unique water culture.

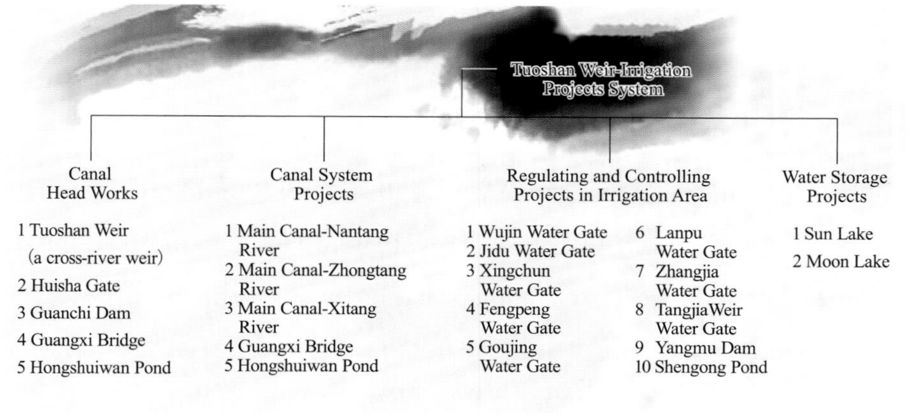

Fig.4 Tuoshan Weir-Irrigation Projects System

VII

Zhidanyuan Sluice of the Yuan Dynasty — one of the best-preserved ancient water projects in China

Built in the Yuan Dynasty (A. D. 1206—1368), Zhidanyuan Sluice boasts of a history of 700 years. This ancient sluice came to light by means of archaeological excavation, standing as one of the best-preserved ancient water projects in China so far. The sluice lies at the junction of Zhidan Road and Yanchang West Road in the Shanghai's Putuo District.

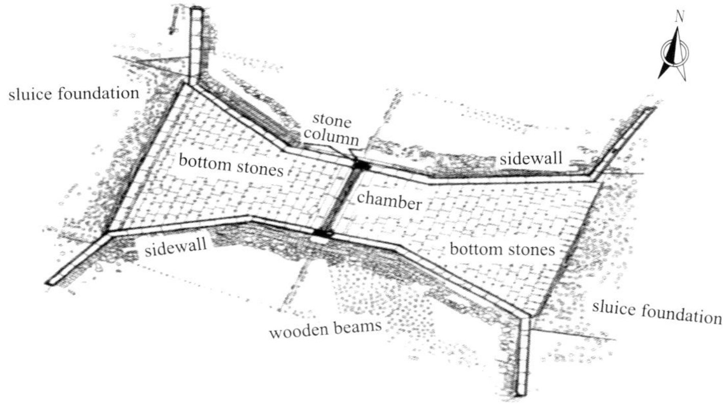

Fig.1 The layout of Zhidanyuan Sluice

In the Qin (221B. C.—206B. C.) and Tang (A. D. 618—907) Dynasties, the Dongjiang River, the Songjiang River and the Loujiang River acted as the main waterways for the Taihu Lake to empty itself into the sea. However, due to the daily tidal waves, these three river courses were easily choked by the sea silts. In the Northern Song

Dynasty (A. D. 960—1127), the Wusong River (the Songjiang River) was the only existing waterway that still connected the Taihu Lake and the sea, but it also became increasingly silts-clogged. Just as the history recorded "two rivers (the Dongjiang River and the Loujiang River) have been clogged, and only the Wusong River (the Songjiang River) remains". Apparently, the siltation of the Wusong River became more serious in the Yuan Dynasty after experiencing natural siltation, artificial dredging and river course shifting. Thus, Zhidanyuan Sluice, a man-made dredging project, was built up as a tailor-made solution to defy the natural siltation of the Wusong River.

Zhidanyuan Sluice was one of the numerous projects built in the Yuan Dynasty to tackle the siltation of the Wusong River. These diverse projects consisted of stone and wooden sluices, whose main roles were to block and clean up tidal sands and silts.

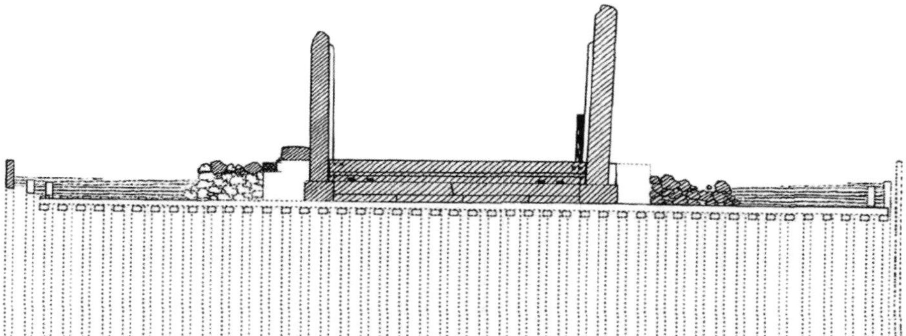

Fig.2 The cross section of Zhidanyuan Sluice

With a width of 6.8 m and a total area of 1,500 m^2, the sluice is composed of gates, walls, bottom stones, rammed earth, etc. Moreover, it features in the sphere of the rigorous layout, elaborate construction method, high-quality materials and outstanding workmanship.

The first step of construction was to select the sluice location.

Once the bottom trench was dug, about 10,000 pine piles with a length of 4–6 m each were staked down to reinforce the soil base. Every pile was numbered to enable each engineering detail to be well recorded. This approach not only shows a monitoring method of the construction process, but also embodies the essence of Zhidanyuan Sluice.

Secondly, the gaps among the piles were filled with gravels and further compacted. Afterwards, place wooden beams on the piles, and lay down stone-lined wooden boards. On top of that, bluestone slabs were placed, serving as bottom stones. Then, approximately 400 iron ingots were deployed to hold these stone slabs together.

Consequently, set up gate walls and then erect gate stone columns. The foundation of gate walls were made of large stones and built on stone-lined wooden boards. Then, multi-layer stone bars were placed on top of those boards. The gate stone columns were sandwiched between these stone walls.

Fig.3 Pine piles

Fig.4 Bottom stones

Finally, outside the stone walls, bricks were lined, and then waste rocks were piled up and filled with lime-soils.

In accordance with archaeological findings, Zhiduanyuan Sluice pursues excellence in the sphere of design, construction, function, workmanship and even gaps among rocks. The quality of construction 700 years ago was not inferior to today's level.

Zhidanyuan Sluice serves as a rare real-world exhibit that sheds light on the history of water engineering in the Wusong River, the Taihu Lake Basin and China as well as the economic strength of Shanghai during the Yuan Dynasty. It is also invaluable material to research the evolution history of Shanghai from town to metropolis. Hence, this facility plays an extremely pivotal role in the history of water engineering and urban development in China.

VIII

Fushou Ditch System — a great contributor to an ancient flood-proof city

Ganzhou City lies in the south of Jiangxi Province where the Zhangjiang River and the Gongjiang River converge. Being surrounded by waters in three sides, the whole city is confronted with a heavy pressure of flood. The Fushou Ditch System, commonly known as the "Fish Intestine Ditch", is an underground drainage system of the ancient city of Ganzhou, with a total length of 12.6 km. During the Northern Song Dynasty (A. D. 960—1127), on the basis of predecessors'

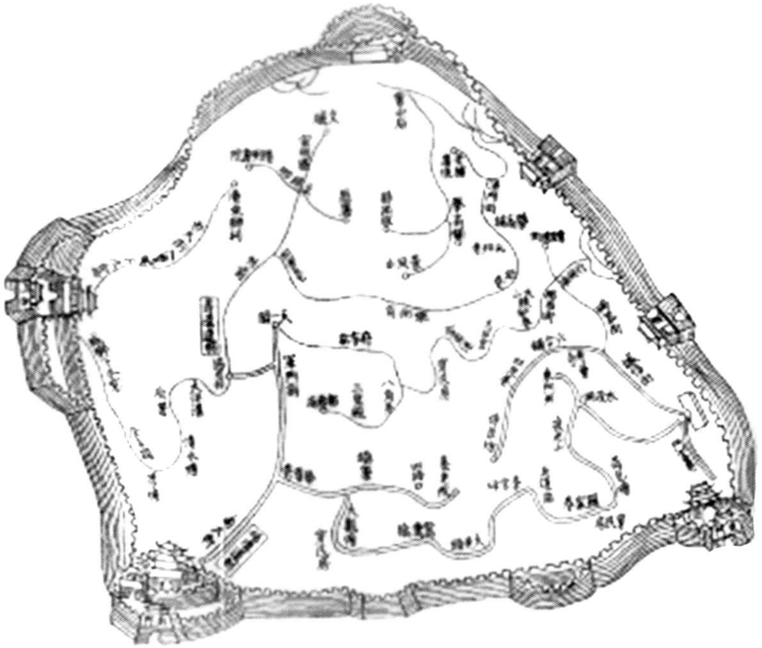

Fig.1 The Fushou Ditch System

endeavors and according to terrain features of Ganzhou City, Liu Yi, the mayor of Ganzhou City, built two underground ditches by the principle of zoning drainage, enabling rainwater and sewage in the city to naturally flow into the ditches and then be diverted to the Zhangjiang River and the Gongjiang River outside the city from the east and west directions.

In 1077, the Fushou Ditch System was completed after 10 years of construction, finally ending flood disasters in Ganzhou City. Local people thus protected their properties and gained happiness and longevity. Meanwhile, because the outlook of two ditches looks like "Fu" (good fortune) and "Shou" (longevity) in the seal script of the Chinese characters, the local people vividly name the drainage system as the "Fushou Ditch System", which not only describes the appearance of the ditches but also conveys sound cultural implications. The Fushou Ditch System boasts of the following characteristics in the sphere of design and construction:

(1) In the field of the craft, the Fushou Ditch System adopts the brick-arched structure, because it enjoys the lowest cost but the best gravitational force. Moreover, this structure enhances the bearing capacity of the ditch wall and prolongs the service life. The arched structure and the walls made of green bricks and hemp stones are intertwined to ensure the long service life of this underground waterway.

(2) Natural imitation is one of the architectural characteristics of the Fushou Ditch System. Ganzhou City is located in a hilly region and the land form inside the city is undulating. Hence, the Fushou Ditch System follows its topography to employ different heights, widths and slopes, which is conducive to both speeding up water flow to rapidly discharge rainwater and flushing sediments in the Ditch to ensure

smooth flow.

(3) The drainage system of the Fushou Ditch System undertakes two major functions of storage during rainstorm and sewage discharge in normal times. Diverse drainage facilities are distributed in residential courtyards and kitchens. Consequently, sewage can be directly discharged from residents' homes to the Ditch and then flow outside the city.

(4) The bottom of the Fushou Ditch System is paved with stones, which has the effect of resisting acid, alkali, heat and corrosion in sewage.

(5) The inscriptive bricks in the basement of the Fushou Ditch System record the names of the craftsmen, serving as a quality traceability system to ensure the project quality. Although the bricks have lasted for nearly 1,000 years, the inscriptions are still legible, proving outstanding

Fig.2 Masonry arch bridge

craft level and craftsmanship of ancient people.

(6) The concept of the Fushou Ditch System bears in mind the maximization of ecological benefits. The Fushou Ditch System is integrated with three big ponds and dozens of small ponds in the city, realizing comprehensive effects of water storage, fish farming, irrigation and sewage treatment. Furthermore, the ecological and environmental protection circulation chain formed thereby coincides with the present concept of sponge city.

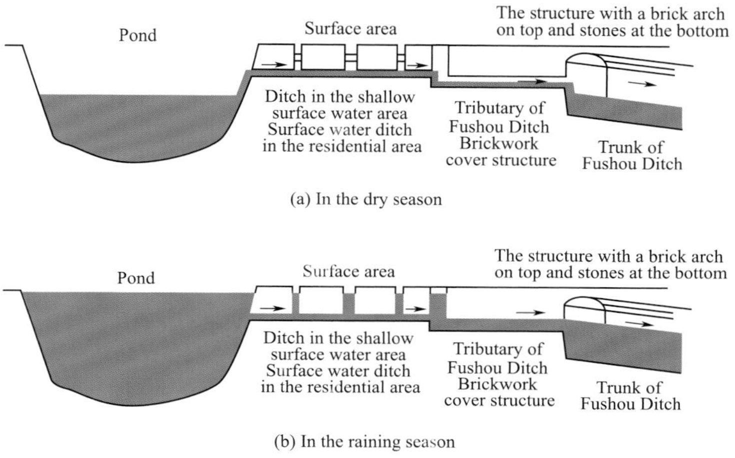

Fig.3 Schematic diagram of drainage of the Fushou Ditch System

The Fushou Ditch System has by far experienced a history of about 1,000 years. Thanks to its scientific, rational and practical architectural form, Ganzhou City takes advantage of water resources without suffering from floods for thousands of years. Besides, the Fushou Ditch System, still in use now, draws the world's attention as a model of the underground drainage system of ancient Chinese cities. Meanwhile, its ingenious water system design still sheds light on today's urban planning and construction.

IX

Jiangxi Weir — Longyou's equivalent of Dujiangyan

With a history of over 600 years, Jiangxi Weir is located in Houtianpu Village at the downstream of Lingshan River in Longyou County, Zhejiang Province. Known as "Longyou's equivalent of Dujiangyan" (Dujiangyan is the oldest functioning dam-free water diversion project in the world), it enshrines the concept of harmony between man and nature, and reveals the wisdom of sustainable irrigation. In 2018, Jiangxi Weir was inscribed on the World Irrigation Project Heritage List.

Lingshan River is the largest tributary of Qu River in Longyou County. As a typical mountain river, it has a winding course, drastically rising and falling down its path. Consequently, it is difficult to use the river for irrigation purpose. Moreover, in the flood season, river water

Fig.1 Layout map of Jiangxi Weir

Fig.2 Current situation of Jiangxi Weir

tends to flood surrounding areas, causing heavy damage to farmland and seriously affecting local production and livelihood. From 1330 to 1333, in order to divert water for irrigation and resist floods, local people took advantage of the sandbar in the river to design and build two weirs. Charkoma (a Mongolian), the magistrate of Longyou County at that time, presided over the project construction and entrusted the Jiang and Xi families to undertake the construction of two weirs respectively. Therefore, the whole project was named as Jiangxi Weir.

By making full use of natural conditions, including the sandbar in the river, natural river course and disparity in river bed height, local people took the sandbar as a link to connect Jiang Weir in the upper stream and Xi Weir in the downstream. Eventually, a pivotal water project was formed by means of integrating water diversion, disaster alleviation, agriculture irrigation, water discharge and sand drainage. Standing in the major course of Lingshan River, Jiang Weir, the main weir, has the function of raising water level, thus facilitating part

of river water to enter the diversion channel along the tributary for irrigation purpose. In the wet season, the river water exceeding Jiang Weir flows directly into the downstream of Lingshan River. Being a tributary weir, Xi Weir aims to discharge overflowing channel water and enable it to return to the lower reaches of Lingshan River.

Based on the pine and pebble framework, the body of Jiangxi Weir appears square in shape. Without mechanical tools, people in ancient time used local materials and wisely combined pines and pebbles to form a solid foundation for the Weir. Thanks to this engineering ingenuity, Jiangxi Weir has been standing upright for hundreds of years and playing a vital role in irrigation until nowadays. Currently, it waters 2,330 hm^2 of farmland. According to the historical record, however, during the reign of Emperor Kangxi (A. D. 1662—1722) of the Qing Dynasty (A. D. 1616—1911), it once irrigated more than 3,340 hm^2 of land, enormously boosting the development and prosperity of local agriculture, transportation, commerce, culture and other industries.

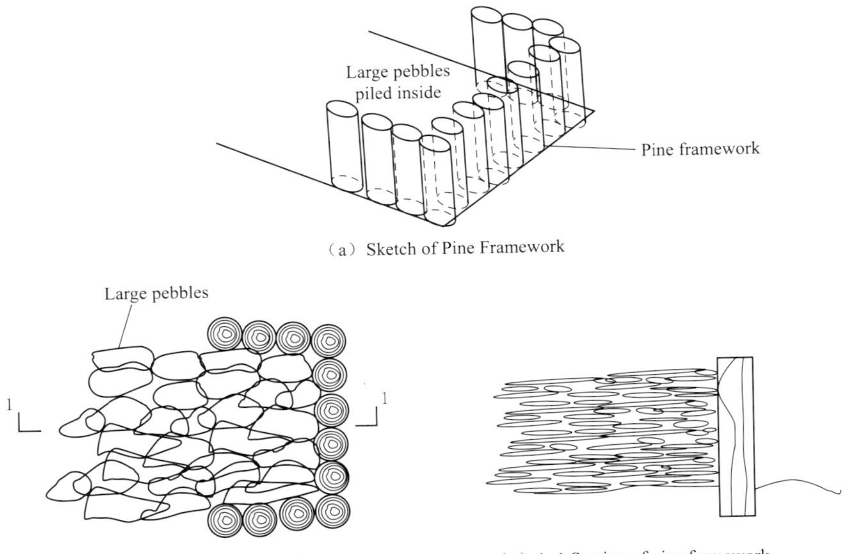

(a) Sketch of Pine Framework

(b) Plan of pine framework

(c) 1-1 Section of pine framework

Fig.3 Pine and pebble foundation

X

Liujiaxia Hydraulic Complex — pearl of the Yellow River

Located in Yongjing County, Gansu Province, Liujiaxia Hydraulic Complex is a large-scale multipurpose water project on the main stream of the Yellow River and the 7th cascade hydropower station in the development plan of the upper reaches of the Yellow River. Initiated in September 1958 and completed in 1974, it is a large hydropower project designed and constructed independently by China during the period of the 1st Five-Year Plan (A. D. 1953—1957), with a total investment of 638 million yuan. Liujiaxia Hydraulic Complex is mainly designed for power generation, with such supplementary benefits as flood control, irrigation, ice-jam prevention, navigation,

Fig.1 Liujiaxia Hydraulic Complex

aquaculture, etc. After completion, it became the largest hydraulic complex in China at that time and was known as the "Pearl of the Yellow River". In November 2018, Liujiaxia Hydraulic Complex was inscribed on the National Industrial Heritage List.

1. Power generation

Five hydroelectric units are totally installed in Liujiaxia Hydropower Station with an installed capacity of 1.225 million kW, and its first 225,000 kW unit was put into operation in March 1969. As China's first megawatt hydropower station, Liujiaxia Hydropower Station is capable of generating 5.7 billion kWh annually. Its power plant is about 25 m wide, 180 m long and 20 stories high, with 5 large China-made turbines at the center, supplying power to Shaanxi, Gansu, Qinghai and other provinces. Mainly undertaking the tasks of power generation,

Fig.2 Liujiaxia Hydropower Project

Fig.3 Liujiaxia Reservoir

peak regulation, frequency regulation and voltage regulation, Liujiaxia Hydropower Station is a backbone power station and plays a pivotal role in the northwestern power grid of China.

2. Irrigation

Liujiaxia Reservoir records a total capacity of 5.7 billion m^3, a controlled watershed area of 173,000 km^2 and an average annual flow of 834 m^3/s. The dam is a concrete gravity one with a height of 147 m, a length of 204 m and a top width of 16 m. Every year, Liujiaxia Reservoir supplies 800 million m^3 of water for spring irrigation in Gansu, Ningxia and Inner Mongolia. The irrigation guarantee rate increases from 65% to 85%, and the irrigation area rises from 670,000 hm^2 to 1.07 million hm^2.

3. Flood control

Liujiaxia Reservoir improves the flood control standard of downstream cascade hydropower stations and Lanzhou City, so that the standard of 1000−year return period of downstream Yanguoxia Hydropower Station is increased to 2000−year return period, and the peak discharge of 100−year return period of Lanzhou City is reduced from 8,080 m^3/s to 6,500 m^3/s.

4. Ice-jam prevention

Ice-jam disaster is a natural disaster that has hit the Yellow River for many years. In the spring-thawing period, water swells, ice breaks and floating ice blocks dams, thus causing serious ice-jam disasters characterized by river flooding and dike failure. After Liujiaxia Reservoir was put into use, about 700 km of the downstream area were protected from ice-jam hazards, and no major ice-jam disasters took place in the past 20 years.

5. Water supply

After completion, Liujiaxia Reservoir meets industrial and urban water demands of Lanzhou City, Yinchuan City and other downstream cities, rendering about 700,000 m^3 of industrial water to Lanzhou City every day.

XI

Baihetan Hydropower Station — six technical indicators ranking first in the world

Baihetan Hydropower Station, located at the junction of Ningnan County, Liangshan Prefecture, Sichuan Province, and Qiaojia County, Zhaotong Prefecture, Yunnan Province, is the second cascaded hydropower station developed on the main stream of the lower reaches of Jinsha River, with such comprehensive benefits as power generation, flood control, sediment retaining, navigation and irrigation.

Fig.1 Baihetan Hydropower Station

Currently, Baihetan Hydropower Station boasts of the largest capacity of per hydroelectric unit. The normal water level of its reservoir is 825 m, with a total storage capacity of 20.6 billion m^3. Its underground powerhouse is equipped with 16 hydroelectric units, with a gross installed capacity of 16 million kW and an average annual power generation of 62.443 billion kWh. The static investment of the project stands at 143.07 billion RMB. The construction of the main body of the project was initiated in 2013, with the first generating unit commissioned in July 2021. The project is slated for full completion in December 2022. After completion, Baihetan Hydropower Station will become the 2nd largest in China after the Three Gorges Hydropower Station.

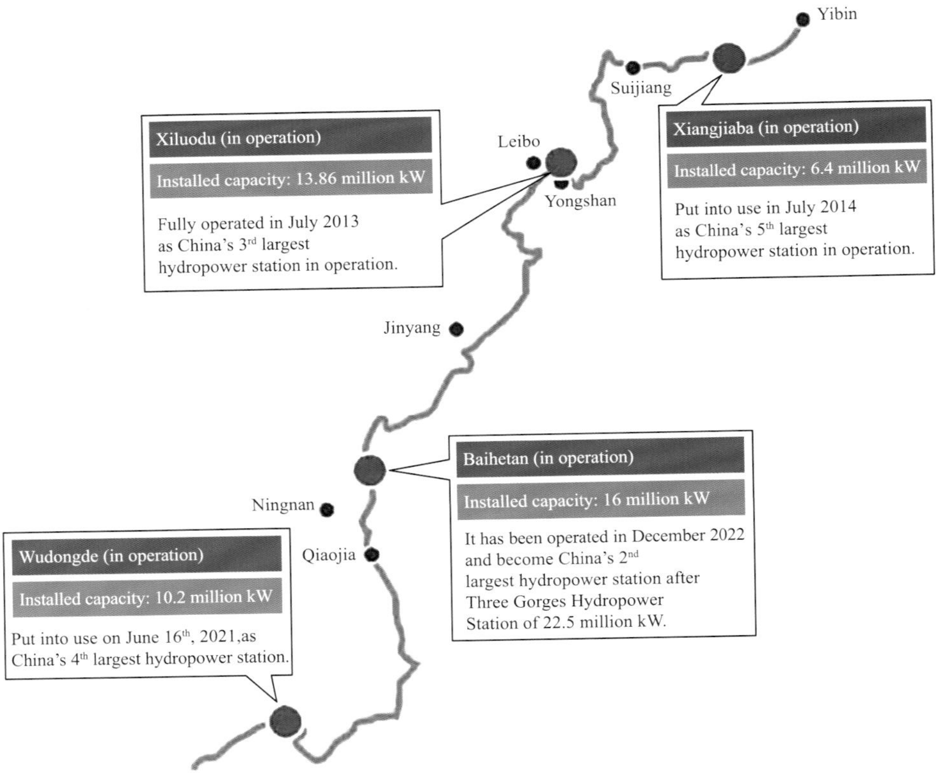

Fig.2 Jinsha River hydropower base

Frg.3 Hydro-turbine generating units under construction

As a backbone project on Jinsha River, Baihetan Hydropower Station is the 2nd cascade among the four hydropower cascades (Wudongde, Baihetan, Xiluodu and Xiangjiaba) in the lower reaches of Jinsha River. It will join the ranks of Three Gorges Hydropower Station and Xiluodu Hydropower Station as a 10-million-kilowatt-class hydropower project. Baihetan Hydropower Station is about 182 km away from Wudongde dam site and 195 km away from Xiluodu dam site, with a controlled catchment area of 430,300 km^2.

During the construction process, six technical indicators of Baihetan Hydropower Station rank first in the world, including the capacity of single hydro-turbine generating unit, the scale of underground cavern groups, the size of the cylindrical tailrace surge chamber, the seismic-resistence parameters of a 300-meter high arch dam, the scale of pressure-free flood discharging tunnel groups, and the adoption rate of low-heat concrete throughout the whole dam body.

Baihetan Hydropower Station is composed of a dam, flood discharge and energy dissipation facilities, water diversion and power generation systems as well as other structures. The dam is a concrete double-curved arch one, with the dam height of 289 m. The arch crest

Fig.4 Baihetan Hydropower Station under construction

thickness is 14 m and the concrete pouring volume of about 8.03 million m^3. Besides, it is fixed with eight China's home-made million-kilowatt-class power units on both left and right banks. As the first batch of million-kilowatt-class power units in the world, they are super-giant mixed-flow hydro-turbine generating units, claimed as the "Mount Everest" of the world's hydropower industry. It is estimated that a single million-kilowatt-class unit in Baihetan weighs more than 8,000 t, equivalent to the weight of the Eiffel Tower in France.

The construction of Baihetan Hydropower Station creates enormous opportunities for local social and economic development, dramatically improving infrastructure situation and boosting related industries in the forthcoming future. Meanwhile, the project has far-reaching impacts on China's west-to-east power transmission program, driving economic development in western China.

XII

Wudongde Hydropower Station — a scientific achievement on the Jinsha River

As a national key project in China's west-to-east power transmission program, Wudongde Hydropower Station, straddling Luquan County, Yunnan Province, and Huidong County, Sichuan Province, is the first among the four cascade hydropower stations (Wudongde, Baihetan, Xiluodu and Xiangjiaba) in the lower reaches of Jinsha River. Equipped with 12 hydro-turbines with a single unit capacity of 850,000 kW, Wudongde Hydropower Station has a gross installed capacity of 10.2 million kW, ranking the 7th in the world. At that time, it enjoyed the largest capacity of single unit in operation in the world. It is also the 4th in China and the 7th in the world that joins the club of mega hydropower stations with over 10 million kilowatts installed capacity.

Wudongde Hydropower Station is composed of main structures for water retaining, water releasing, water diversion and power generation, etc. The water retaining structure is a concrete double-curved arch dam, with the crest elevation of 988 m, the maximum dam height of 270 m and the bottom thickness of 51 m. Its thickness-height ratio is 0.19, ranking the project as the present world's thinnest among arch dams above 300 m in height. Besides, it is the first-ever super-high arch dam poured with low-heat cement concrete throughout the whole dam body in the world. The total storage capacity of Wudongde Reservoir stands at 7.408 billion m^3, with the regulated storage capacity of 3 billion m^3 and the flood control capacity of 2.44 billion m^3. Its powerhouses are

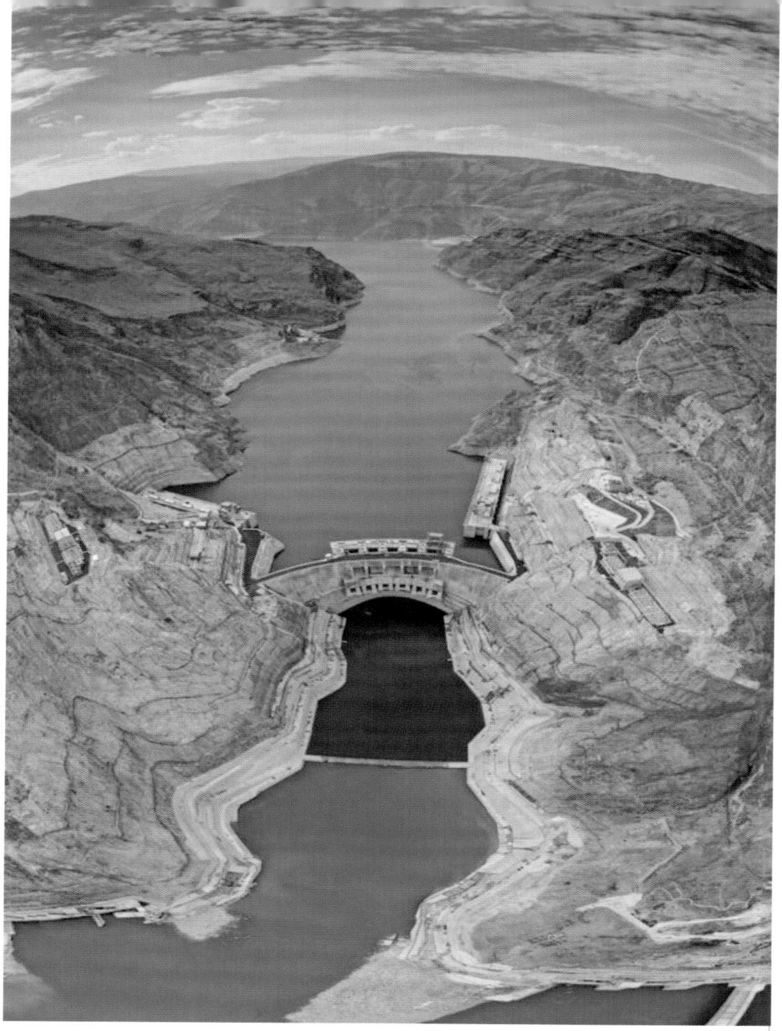

Fig.1 Wudongde Hydropower Station

built in the mountains on both banks, with 6 mixed-flow hydro-turbine generating units installed on each side. The designed annual power generation is 38.91 billion kWh, and the daily power generation will cater for an annual power consumption by 300,000 people.

On December 24, 2015, the construction of Wudongde Hydropower Station was kicked off. On March 16, 2017, the concrete pouring of the dam commenced. On June 29, 2020, the first batch of units came into operation. On June 16, 2021, all units were put into use. At present, Wudongde Hydropower Station has fully reached the level of 945-meter water storage.

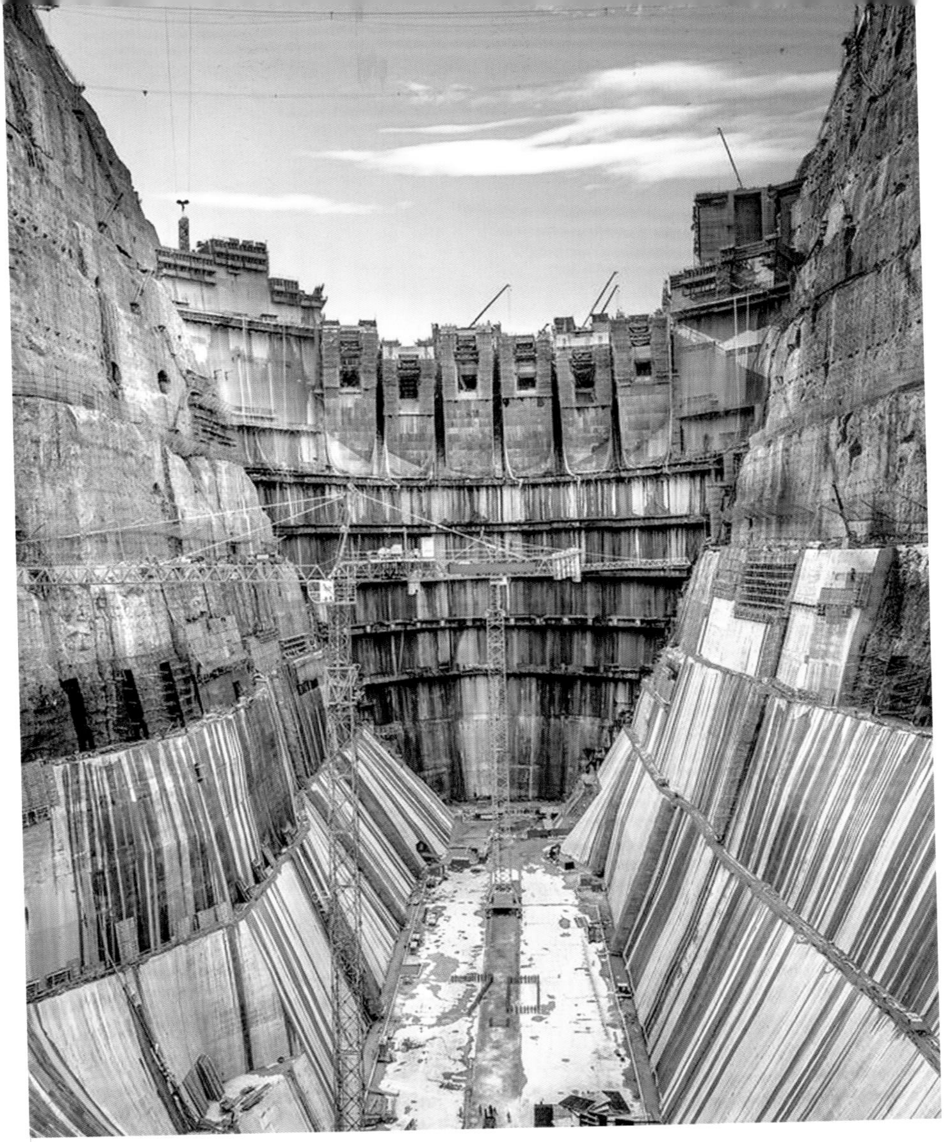

Fig.2 Wudongde Hydropower Station under construction

The four hydropower stations (Wudongde, Baihetan, Xiluodu and Xiangjiaba) in the lower reaches of Jinsha River have a combined storage capacity of 45.868 billion m^3 and a flood control capacity of 15.493 billion m^3. Being a vital component of the flood control system of Yangtze River, they will further step up the flood control capacity of Sichuan and Yunnan provinces.

Wudongde Hydropower Station is a significant cascade project in the development of the Jinsha River basin, with the main function of power generation and the supplementary benefits of flood control,

navigation and sediment retaining. Hence, it creates favorable conditions for power generation and economic benefits of downstream hydropower stations, and promotes local poverty alleviation, contributing to local economic and social development.

Tab.1 Top 10 hydropower stations in China

No.	Power station	River Basin	Installed capacity /10,000 kW	Annual power output /billion kWh
1	Three Gorges	The upper reaches of Yangtze River	2,250	882
2	Baihetan	Jinsha River	1,600	624
3	Xiluodu	Jinsha River	1,260	620
4	Wudongde	Jinsha River	1,020	389
5	Xiangjiaba	Jinsha River	640	307
6	Nuozhadu	Lancang River	585	239
7	Longtan	Hongshui River	490	157
8	The 2nd cascade of Jinping	Yalong River	480	242
9	Xiaowan	Lancang River	420	190
10	Laxiwa	The upper reaches of Yellow River	420	102
Total			9,165	3,752

XIII

Xiluodu Hydropower Station — a national key project in China's west-to-east power transmission program

Xiluodu Hydropower Station, straddling Leibo County, Sichuan Province, and Yongshan County, Yunnan Province, is the third among the four cascade hydropower stations (Wudongde, Baihetan, Xiluodu and Xiangjiaba) in the lower reaches of Jinsha River. It boasts of comprehensive benefits such as power generation, flood control, sediment retaining, navigation improvement for the upper reaches, etc. Moreover, it is not

Fig.1 Xiluodu Hydropower Station

Fig.2 The double-curved arch dam

only the largest hydropower station in operation in Jinsha River, but also a national key project in China's west-to-east power transmission program. In September 2016, Xiluodu Hydropower Station won the "FIDIC 2016 Engineering Project Excellence Award".

In April 2007, the construction of Xiluodu Hydropower Station was kicked off. The project is composed of main structures for water retaining, water releasing, water diversion and power generation, etc. This mega-project is famous for its high dam (285.5 m), high seismic resistance capacity (Ⅷ degree) and high-speed water flow (close to 50 m/s), as well as large discharge (maximum discharge of about 50,000 m^3/s), large underground powerhouse (roof-arch span of over 30 m) and large-scale hydro-turbine generating units (a single unit capacity of 770,000 kW).

As the third highest in China, the dam of Xiluodu Hydropower Station is a concrete double-curved arch one, with the height of 285.5 m and the concrete pouring volume of about 6.8 million m^3. Equipped with nine giant hydro-turbines with a single unit capacity of 770,000 kW

on both left and right banks, Xiluodu Hydropower Station boasts of a total installed capacity of 13.86 million kW and an average annual power generation of 57.12 billion kWh. Currently, it ranks the third largest hydropower station in operation in China and the fourth in the world.

With the flood control capacity of 4.65 billion m^3, the total storage capacity of Xiluodu Reservoir stands at 12.67 billion m^3. Hence, it becomes a pivotal component of the Yangtze River flood control system and one of the main engineering measures to tackle the flooding problem in the Sichuan section of the Yangtze River, upgrading the flood control standard of Yibin, Luzhou, Chongqing and other cities along the river from once in 20 years to once in 100 years.

Fig.3 Xiluodu Reservoir

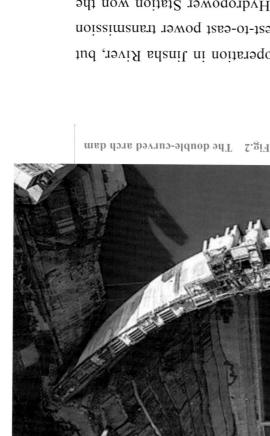

Xiluodu Hydropower Station is of epoch-making significance in realizing rational distribution of energy, optimizing power supply structure, improving ecological environment, stepping up economic development of the western region, especially the minority areas on both sides of Jinsha River in Sichuan and Yunnan, and boosting the sustainable social and economic development of the Yangtze River Basin.

XIV

Xiangjiaba Hydropower Station — the 5th largest Hydropower Station in China

Xiangjiaba Hydropower Station, straddling Zhaotong City, Yunnan Province and Yibin City, Sichuan Province, is the last among the four cascade hydropower stations (Wudongde, Baihetan, Xiluodu and Xiangjiaba) in the lower reaches of Jinsha River. Moreover, it ranks as the 5th largest hydropower station in China and the 11th in the world, and also functions as the backbone power source for

Fig.1 Xiangjiaba Hydropower Station

Fig.2 Xiangjiaba Reservior

China's west-to-east power transmission program. The construction of Xiangjiaba Hydropower Station was commenced in November 2006, and it was fully commissioned in July 2014.

With a total static investment of about 54.2 billion RMB, Xiangjiaba Hydropower Station is mainly composed of a main dam, an underground powerhouse on the right bank, a powerhouse behind the dam on the left bank, navigation buildings, irrigation water intakes, etc. The dam is a concrete gravity one with a height of 162 m and a crest length of 909.26 m. Xiangjiaba Reservoir is a canyon-typed one with a total storage capacity of 5.163 billion m^3 and a control catchment area of 458,800 km^2, accounting for 97% of the Jinsha River Basin. Furthermore, equipped with eight 800,000 kW turbines and three 450,000 kW turbines, Xiangjiaba Hydropower Station possesses a total installed capacity of 7.75 million kW and an average annual power generation output of 30.747 billion kWh.

Xiangjiaba Hydropower Station is the only station fixed with a ship lift and enjoying an irrigation function among 25 hydropower stations in the Jinsha River Hydropower Base. Besides, it is capable of tackling

the biggest problem with which the Three Gorges Project has been confronted, sediment deposition. The middle reaches of Jinsha River is one of the main sediment source areas of the Yangtze River, making up about 1/2 of the sediment inflow of the Three Gorges Reservoir. After the completion of Xiangjiaba Hydropower Station, the sediment migration into the Three Gorges Reservoir has been eliminated by more than 34%.

During the construction process, six technical indicators of Xiangjiaba Hydropower Station rank first in the world, including the scale of open caisson group, the capacity of single hydro-turbine generating unit, the size of single ship lift, the scale of flood stilling basin, the span of cable crane and the length of sand and gravel conveyor belt.

Aside from power generation, Xiangjiaba Hydropower Station also boasts of comprehensive benefits such as improving navigation conditions, flood defense, irrigation, sediment detention, reverse regulation, etc. Hence, it creates sound opportunities for local social and economic development, and provides a tremendous booster to regional infrastructure conditions.

Fig.3 Ship lift

XV

Sanmenxia Project—a mainstay project on the Yellow River

Fig.1 Sanmenxia project dam

Sanmenxia project is located in the lower section of the middle reaches of the Yellow River. Renowned as the "Mainstay Project on the Yellow River", it is the first large-scale water project built on the main stream of the Yellow River.

The construction of Sanmenxia project started in 1957, and it was put into operation in September 1960. The total length of the main dam is 713.2 m, and the maximum dam height is 106 m. The dam crest elevation is 353 m. With a flood control capacity of nearly 6 billion m^3, its

Fig.2 Aerial view of Sanmenxia project

operation can cover 688,400 km^2 of the Yellow River basin, accounting for 91.5% of the basin area. Its design was originally entrusted to Leningrad Design Bureau of the Soviet Union. The reconstruction design was undertaken by Tianjin Survey Design Institute (China Water Resources Beifang Investigation, Design and Research Co., Ltd at present), and its construction is carried out by the 11th Engineering Bureau of the former Ministry of Water Resources and Electric Power.

Since its completion and operation, Sanmenxia project has given full play to the comprehensive benefits of flood control, ice prevention, water and sediment regulation, irrigation water supply and power generation. It not only serves as an important part of the flood control and silt reduction project system in the lower reaches of the Yellow River, but also promotes the all-round development of local economy and society. The

sediment concentration of the Yellow River decreases year by year. Ice floods and dam bursts have been prevented in the lower reaches of the Yellow River. Areas irrigated by water from the Yellow River increased tremendously. Water demand of 20 prefectures, cities and more than 100 counties along the Yellow River can be fully satisfied. The sediment in the lower reaches of the Yellow River can be discharged into the sea smoothly.

At present, more than 200 km^2 of waters in the reservoir area has become a national wetland nature reserve, which is essential for regulating the regional weather, protecting local biodiversity and improving the ecological environment. As one of the two important large hydropower stations in Henan power grid, Sanmenxia hydropower station has an installed capacity of 450 MW. From 1973 to the end of 2020, it has provided more than 56 billion kWh of green power to the power grid, equivalent to reducing coal combustion by about 16 million t, sulfur dioxide emissions by about 7,900 t and soot emissions by about 44,400 t. Therefore, it has made positive contributions to urban development and ecological environment protection.

Fig.3 Water and sediment regulation of Sanmenxia project

XVI

The Red Flag Canal — the 8th wonder of the world

Known as "the Great Wall of Water", the Red Flag Canal is a water project built by the people of Linzhou, Henan Province from the 1960s to 1970s. It was built on the side of cliffs of the Taihang Mountain to divert water from Zhanghe River to the Linzhou County.

Fig.1 The Red Flag Canal

Fig.2 The construction site

The construction of the Red Flag Canal commenced in February 1960. A total of 300,000 people of Linzhou were mobilized to undertake the engineering work on the top of the Taihang Mountain under extremely harsh conditions for ten years. The industrious Linzhou people leveled 1,250 hilltops, erected 151 aqueducts, dug 211 tunnels, built 12,408 hydraulic structures, and excavated 22.25 million m^3 of earth and rock, accomplishing a feat of water engineering.

Taking the Zhuozhang River as its source of water, the Red Flag Canal extends 1,500 km in length. The main canal wall is 4.3 m high, 8 m wide and 70.6 km long. A total of 45 small hydropower stations were built by taking advantage of the high water head of the Red Flag Canal. Thanks to its multiple functions of water diversion, storage, lifting, drainage, and power generation, the Canal has irrigated a large agriculture area and stood out as a scenic spot. The construction of the

Canal not only ended the miserable history of "nine drought years out of ten and water as expensive as oil" in the Linzhou County, but also fundamentally improved people's living conditions. The people of Linzhou affectionately call the Red Flag Canal "the Canal of life" and "the Canal of happiness".

The Red Flag Canal has become a glorious banner of water engineering in China. Former Premier Zhou Enlai once proudly told foreign guests: "There are two wonders in the newly founded People's Republic of China. One is the Nanjing Yangtze River Bridge and the other is the Red Flag Canal in Linzhou". The Canal is deemed by many visitors as a magnificent achievement, and was recognized as a national key cultural heritage protection site by the State Council in 2006. Nowadays, the influence of the Canal is far beyond a water project. The spirit of "self-reliance, hard work, solidarity, cooperation and selfless devotion", as embodied by Linzhou people in building the Red Flag Canal, has become an epitome of the Chinese psyche.

Fig.3 The Red Flag Canal

XVII

Changqu Canal (Bai Qi Canal) — a project constructed two thousand years ago due to war

Fig.1 Changqu Canal

In the city of Xiangyang in China, there is a 2000-plus-year-old irrigation system where reservoirs and ponds are like melons, and canals are like vines. Because the main canal ran over 50 km, this irrigation system is called the Changqu (meaning "long canal" in Chinese).

Bai Qi, a general of the Qin Dynasty (221 B. C.—206 B. C.), had the weirs and canals built in 279 B. C. and used the water to inundate the enemy troops. Once built for war, this project later became an important irrigation system in the Xiangyang Plain. Changqu Canal was also named the Bai Qi Canal to commemorate its original creator.

Fig.2 General Bai Qi

In the 3rd century, the Muqu Canal was built to the north of the Changqu Canal, which diverted water from the Yanshui River, another tributary of the Han River. The Changqu Canal and the Muqu Canal were two canal systems connected with each other. A large irrigation district with 40,000 hm^2 was developed at today's Xiangyang and

Fig.3 The connected Changqu and Muqu Canals

Fig.4 The melons-on-vines irrigation system

Yicheng. The Xiangyang-Yicheng Plains became a "bread basket" in the middle reach of the Han River.

After over 600 years of gradual improvement from the Qin and Han Dynasties until the Southern and Northern Dynasties (A. D. 420—589), the Changqu and Muqu Canals had already become a regional irrigation project with water diversion and storage functions by the 5th century at the latest. From the 13th century to the early 20th century, the Changqu and Muqu Canals had always been managed under the system of governmental supervision and private management. However, the canals fell into disrepair in the 19th century because of the sharp conflict between the needs of upstream water transport and downstream irrigation. They were finally rebuilt at the original sites in the mid-20th century.

The Changqu Canal irrigation system consists of headworks, canal system as well as regulation and storage works. The headworks are located at the confluence of the main stream of the Manhe River and its tributary Qinglianghe River near Wu'an Town in Nanzhang County. The main canal runs through the middle of the irrigation district.

At present, there are 15 medium and small sized reservoirs and 2,671 weirs and ponds in the irrigation district. The reservoirs are connected to the Changqu Canal through ditches and canals controlled by water gates. The Changqu Canal is like a vine, and the reservoirs, weirs and ponds connected to it are like melons. The project has become

Fig.5 Changqu Canal

a typical melons-on-vines irrigation system in South China. During non-irrigation season, barrages divert river water into canals and thereby feed into reservoirs and ponds. During the irrigation season, the reservoirs and ponds supply water to the Changqu Canal. The repetition of this cycle raises the utilization rate of the reservoirs and ponds and guarantees water supply for irrigation.

The sustained use of the Changqu Canal for two thousand years has made Xiangyang a fertile land and has created a rich and diversified local culture. Memorial temples have been built along the canal to worship people who contributed to the construction and maintenance of the canal, so as to encourage later generations to build irrigation project and develop agriculture. The Changqu Canal still tells its time-honored legend nowadays, even though thousands of years have elapsed.

XVIII

Danjiangkou Reservoir—Asia's Heavenly Lake

Sitting in the upper and middle reaches of the Han River, a tributary of the Yangtze River, Danjiangkou Reservoir straddles Hubei Province and Henan Province. Renowned as a Heavenly Lake, the Reservoir is the largest artificial freshwater lake in Asia as well as the main water source for the middle route of the South-to-North Water Diversion Project.

The Reservoir was eventually formed in 2013 after heightening

Fig.1 Danjiangkou Reservoir

Fig.2 Danjiangkou Dam

the Danjiangkou Dam which was constructed in 1958. With a total surface area of 864 km^2, it controls a drainage area of 1,022.75 km^2 and possesses a storage capacity of 29.05 billion m^3. Its normal water level hits 170 m and its average annual water inflow stands at 39.48 billion m^3. Danjiangkou Water Project is mainly composed of a dam, a hydropower plant, ship lifts and two irrigation canals. The dam height is 162 m, while its length reaches 2,494 m. The power plant is equipped with 6 power units, generating an annual electricity output of 4.5 billion kWh on average.

The Reservoir is a key component of the middle route of the South-to-North Water Diversion Project. The clean water diverted from the Reservoir runs across the Yellow River, and eventually supplies over 20 large and medium-sized cities in Henan and Hebei provinces as well as Beijing and Tianjin municipalities, effectively alleviating the thirst for water in dry North China plain.

Apart from water supply, Danjiangkou Reservoir also generates such comprehensive benefits as flood control, power generation, navigation, and irrigation, etc. It relieves the flood risk for 23 cities including Wuhan and Xiangyang and more than 1.24 million hm^2 of farmland, and improves inland navigation in Hubei and Henan provinces. Danjiangkou Reservoir irrigates over 240,000 hm^2 of farmland and creates favourable conditions for freshwater fish farming. Furthermore, the booming tourism industry surrounding the Reservoir has made important contribution to local social and economic development.

Fig.3 Danjiangkou Dam

Fig.4 Danjiangkou Reservoir

XIX

Xinghua Duotian Irrigation and Drainage System — inscribed on two global heritage lists

Xinghua Duotian Irrigation and Drainage System is located in the hinterland of the Lixiahe Plain on the eastern coast of Jiangsu Province, China, a region with a dense network of rivers and abundant water resources. More than 7,000 years ago, during the Neolithic period, the Lixiahe plain was a gulf between the Yangtze River and the Huaihe River. As the coastline moved eastward, this area formed into an enclosed lagoon and then gradually into a plain.

From 1194 to 1855, the Yellow River encroached on the lower reaches of the Huaihe River to enter the sea, bringing frequent flooding and large amounts of sediment to the Lixiahe area and silting up the local

Fig.1 The aerial view of Xinghua Duotian Irrigation and Drainage System

lakes and swamps. During this period, in response to the frequent flooding, local people dug canals in shallow water and mounded the earth into elevated platforms. After years of hard work, these raised fields were gradually transformed into arable land, called Duotian by the locals.

Duotian is irrigated using bailing buckets. When the water level is high, people stand on boats and irrigate the fields with water directly bailed from the canals. When the water level is low, people water the fields by forming a water-bailing cascade. For the highest field, the cascade would have 4 to 5 stages. Duotian's water level is high in summer and autumn and low in winter and spring. When the water level is low, people raise the height of the fields with the silt scooped out from the canals to prepare for the possible flooding.

After harvests in summer and autumn, people throw the rest of

Fig.2 Farmers working in Xinghua Duotian

Fig.3 Xinghua Duotian

the crops into the canal to produce wet compost. In winter and spring, this natural and organic fertilizer is applied to crops. The abundance of water combined with the high fertility of the soil has enabled the raised fields to produce high-quality fruits and vegetables since ancient times. Vegetable production has become one of the main sources of income for local farmers.

The unique water landscape, sound ecological environment as well as varied folk customs and culture are all valuable resources for the local tourism industry. Famous scenic spots such as the rapeseed Duotian, the forest on water, the wetland park and the fishing eco-park serve as a key driver for the local economy. In October 2022, it was added to the World Irrigation Project Heritage List. In April 2014, Xinhua Duotian Irrigation and Drainage System was officially recognized by the Food and Agriculture Organization of the United Nations (FAO) as a Globally Important Agricultural Heritage.

XX

The Shangbao Terraces—the largest Hakka terraces

Sitting in the northwest of the mountainous Chongyi County of Jiangxi Province of China, the Shangbao Terraces cover an area of about 3,400 hm^2. It is distributed in 26 villages of the three townships of Shangbao, Fengzhou and Sishun. With an altitude ranging from 1,260 m to 280 m, the terraces register a vertical drop as large as 1,000 m, of which the largest one features as many as 62 levels of ridges. In 2012, it was certified as the title "the largest Hakka terraces in the world" by Shanghai China Records. In April 2018, it was listed

Fig.1 The aerial view of the Shangbao Terraces

Fig.2 The Shangbao Terraces

as a “Globally Important Agricultural Heritage System” by FAO. In October 2022, it was added to the World Irrigation Project Heritage List.

According to *the Classic of Mountains and Seas* (known as *Shanhaijing,* an ancient book about a myth of paradise) and other historic documents, the development of the Shangbao Terraces dates back to the pre-Qin period before 221 B.C. Later in the Qin and Han Dynasties, the terraces gradually took shape. After regular repairs and expansions in the later dynasties of Tang, Song, Yuan, Ming and Qing, the terraces eventually obtained its present magnificence. The remains of the stone dog preserved at the Chishui Village of Shangbao Township shed light on the early history of the terraces.

Shaped by the mountainous terrain, the Shangbao Terraces flourish because of water. The densely forested mountains help store abundant water, and the granite mountain structure constitutes an impermeable

reservoir. Therefore, rainwater harvested on the mountain top can permeate along the hillside slopes, creating a natural underground water distribution and drainage network.

In general, the Shangbao Terraces adopt gravity irrigation. Besides, other irrigation methods are also applied in different water demand scenarios such as isolated hills and seasonal and meteorological changes. The irrigation system of the terraces can be divided into 3 categories, namely the water storage structures, the canal network, and water flow regulation facilities. There are 3 ways of water storage, including water directly seeping out of the mountain slopes, artificial water storage and distribution structures, and man-made filter and sedimentation facilities. Under the category of canal irrigation network, several sub-categories exist, including those networks diverting water with weirs, delivering water through inverted siphons, using the small plots of farmland as canals, delivering water with aqueducts and bamboo tubes, and lifting water with waterwheels. For the purpose of water flow regulation, water distributing stones and pipeline valves are installed.

At present, the structure and layout of this gravity irrigation system remains intact. As a legacy of the hard work and wisdom of the ancient Chinese people, it has become an important part of the local agro-economy and been making further contribution to the development of irrigation agriculture.

XXI

Songgu Irrigation Scheme — an eco-museum

As the largest grain-producing area of Oujiang River Basin in Zhejiang Province, Songgu Irrigation Scheme boasts of an irrigated area of over 11,000 hm^2. As an old Chinese saying goes, when Songyang County has a bumper harvest, the entire region of Chuzhou Municipality will enjoy sufficient food supply, which reflects the abundance and prosperity of Songgu Irrigation Scheme.

Fig.1 One of the projects of Songgu Irrigation Scheme—Wuyang Weir

In 138 B. C., a small kingdom of Dong'ou moved northward and relocated at Gushi Town of Songyang County. Thanks to the excellent local irrigation condition, some soldiers and civilians of Dong'ou started land reclamation, thus transforming Gushi Town into a place with well-developed agriculture and thriving market, and launching Songyang's history of agricultural irrigation. In the period of 1041—1048, Bairen Weir was built on the south bank of Songyinxi Stream at the foot of Dushan Mountain. Two weirs of Fangxi stream were built

Fig.2 Songgu Irrigation Scheme

in the village of Shisandu, irrigating more than 500 hm^2 of farmland. In 1340, Bailong Weir was constructed in the south of Songyang County, watering over 130 hm^2 of cropland. During the Song and Yuan Dynasties, Songgu Irrigation Scheme gradually took shape.

In late Ming and early Qing Dynasties, Songyang County boasted of 116 weirs, ranking first among all counties in Oujiang River Basin. The ancestors of Songyang took local terrain into consideration when building weirs and canals for water diversion, and gradually built Songgu Irrigation Scheme with Songyinxi Stream as its water source. Moreover, it has a complete management system and an irrigation network with densely covered weirs and crisscrossed canals. At present, there exist 14 ancient weirs that can still irrigate hundreds of hectares each in Songgu Irrigation Scheme. Such great density is of extreme rarity.

In the Water Museum of Songyang County, there is a collection of 16 notices, statements and inscriptions left by 14 county magistrates, which record the history of rehabilitation, management, water distribution and water disputes in Songgu Irrigation Scheme over a period of 555 years and underline the spiritual heritage of irrigation agriculture in Songyang.

XXII

Tonghui Canal— the northern-most point of the Grand Canal

Tonghui Canal, also known as Tongji Canal, belongs to the northern water system of the Grand Canal of China. It was called Jinshui Canal in the Yuan Dynasty (A. D. 1206—1368) and renamed Yu Canal after the Ming Dynasty (A. D. 1368—1644). Running from east to west, the canal was originally about 82 km long. Its construction kicked off in the Yuan Dynasty under the leadership of a water expert, Guo Shoujing. The first emperor of the Yuan Dynasty named it Tonghui Canal.

Fig.1 Tonghui Canal in the late Qing Dynasty (A. D. 1616—1911)

Fig.2 The estuary section of Tonghui Canal in the old days

However, the heyday of Tonghui Canal was momentary. With the collapse of the Yuan Dynasty and the establishment of Nanjing as the capital, the canal was once abandoned. It was not until the reign title of Yongle (A. D. 1403—1424) of the Ming Dynasty when the capital moved to Beijing that the governance of the Tonghui Canal commenced again. The Ming Dynasty carried out several large-scale renovations on Tonghui Canal. Except for direct entry into Beijing, the grain transport regained prosperity in the Yuan Dynasty, and the canal continued to be used until the early 20th century.

After the cessation of grain transport, Tonghui Canal undertook the task of delivering irrigation water and became the largest drainage system in urban Beijing. With the industrial development, Tonghui

Fig.3 Tonghui Canal at present

Fig.4 Tonghui River in modern times

Canal became the main channel for sewage discharge in urban Beijing. As the sewage was then discharged into the Gaobeidian sewage treatment plant, the water quality of Tonghui Canal has been improving year by year. Eventually, it has become a landscape in Beijing after decades of management.

Nowadays, with a total length of 21 km, Tonghui Canal runs across the core area of Beijing from the East Second Ring Road to the East Fifth Ring Road, connecting Beijing Railway Station, CBD, cultural and innovation pilot zones and other core functional areas. Its drainage area boasts of 250 km^2 and the maximum drainage capacity sits 350 m^3/s. The main tributaries consist of the moat, urban river system, Jin River, Chang river and Nanhan River.

Tonghui Canal was once an economic lifeline of Beijing and played a crucial role in sustaining the prosperity of the economy. Nowadays, Tonghui Canal is a major waterway connecting Beijing with the administrative sub-center of Tongzhou. In October 2020, the Gaobeidian Lake section of Tonghui Canal was open to boat tour, which is vital to riverine ecological environment and tourism development between Beijing and Tongzhou.

XXIII

Daicun Dam —
the first dam in the ancient China

Sitting in the junction of the Dawen River and Daqing River in Dongping County, Shandong Province, Daicun Dam is recognized as the first dam in the ancient China, equally famous as Dujiangyan Weir (the oldest functioning dam-free water diversion project in the world). It was built in the 9th year of the Yongle reign (A. D. 1411) of the Ming Dynasty with a history of more than 600 years. The main purpose of Daicun Dam is to divert the water of the Dawen River to the Grand Canal and ensure its smooth navigation. The Dam plays an irreplaceable role in national unity, economic development and cultural integration of the Ming and Qing Dynasties and has been acclaimed as the Heart of the Grand Canal. In 2014, Daicun Dam was inscribed on

Fig.1 The aerial view of Daicun Dam

Fig.2 Daicun Dam

the World Heritage List by the 38th World Heritage General Assembly.

In the 9th year of the Yongle reign of the Ming Dynasty, Zhu Di, the Emperor of Ming Dynasty, ordered Song Li, Minister of Public Works, to supervise the reconstruction of the Canal. Song Li recruited more than 165,000 people to build a new 60 km canal. At the beginning, Song Li and other officials were incapable to tackle the frequent drying-up problem of the Canal. Finally, they adopted the advice of Bai Ying, a folk expert on water governance, and built Daicun Dam to prevent the water of the Dawen River from running into the sea. After completion, the water was diverted southward into the Xiaowen River before flowing north and south. From then on, it properly solved the cutoff problem of the Canal in hilly areas and helped the Canal to fulfill its mission as the main artery of water transportation. In addition, when the water was abundant in summer and autumn, Daicun Dam could make the water of the Dawen River overflow into the Daqing River to alleviate flooding. When the water dried up in spring and winter, it could intercept the south-flowing water to prevent the drying up of the Dawen River.

The Dam has a total length of 1,599.5 m and controls a watershed area of 8,537 km^2, accounting for 80% of the total river basin of the Dawen River. Daicun Dam consists of Zhushi Dam, Dougong Dike and Huitu Dam, which are utilized and protected by each other. The ladder-shaped Zhushi Dam is composed of Linglong Dam, Luanshi Dam and Gunshui Dam. Zhushi Dam is stone structures made of boulders weighing tons. They are bonded to each other by the belt-and-buckle method, which adopts iron buckles to lock the dam into a single unit.

Through hundreds of years and after numerous floods, Daicun Dam is still locked with iron buckles, which sheds light on the wisdom of Chinese people in water governance represented by Bai Ying. After visiting Daicun Dam in the early 19th century, Fang Wei, a Dutch water expert, spoke highly of the project as a great achievement in the 14th and 15th centuries. In November 1965, Chairman Mao Zedong praised the Dam as a remarkable project and commended Bai Ying, who designed and led the construction of the project, as a farmer water expert. Daicun Dam enjoys a very high value of research in the aspects of functional role, design philosophy, architectural structure and construction techniques. It is a great miracle in the history of water governance in China.

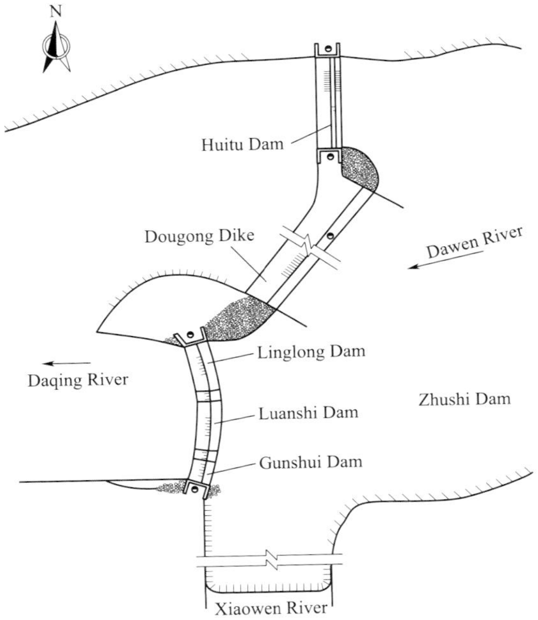

Fig.3 The layout of Daicun Dam